"This ground-breaking volume makes an exceptionally compelling case for the relevance of critical plant studies to youth literature. Drawing from and extending perspectives from ancient Western and contemporary Indigenous philosophy, new materialist and posthumanist thought, and scientific discoveries of the communicative capacities of plants, the authors focus attention on the myriad ways the diverse literature of childhood, with its vivid depictions of sentient vegetal life, orients young readers toward or away from ethical human–plant relations. Engaging with global texts ranging from folklore to contemporary YA fantasies, child-authored poetry to Indigenous stories, picturebooks to middle grade fiction, this is essential reading for anyone interested in environmental humanities, ecopoetics, and children's literature studies."

Karen Coats, *University of Cambridge, UK*

Plants in Children's and Young Adult Literature

From the forests of the tales of the Brothers Grimm to Enid Blyton's *The Faraway Tree*, from the flowers of Cicely Mary Barker's fairies to the treehouse in Andy Griffiths and Terry Denton's popular Treehouse series, trees and other plants have been enduring features of stories for children and young adults. Plants act as gateways to other worlds, as liminal spaces, as markers of permanence and change, and as metonyms of childhood and adolescence. This anthology is the first compilation devoted entirely to analysis of the representation of plants in children's and young adult literatures, reflecting the recent surge of interest in cultural plant studies within the environmental humanities.

Mapping out and presenting an internationally inclusive view of plant representation in texts for children and young adults, the volume includes contributions examining European, American, Australian, and Asian literatures and contributes to the research fields of ecocriticism, critical plant studies, and the study of children's and young adult literatures.

Melanie Duckworth is Associate Professor of English Literature at Østfold University College, Norway, where she teaches British, postcolonial, and children's literature. Her research interests include Australian literature, plant studies, children's literature, and ecocriticism, and she has published on Australian historical children's fiction, Australian literature, ecofeminism, and contemporary poetry.

Lykke Guanio-Uluru is Professor of Literature at Western Norway University and researches literature and ethics, particularly plant studies, ecocriticism, fantasy, and game studies. She is the author of *Ethics and Form in Fantasy Literature* (2015) and multiple research articles, and co-editor of *Ecocritical Perspectives on Children's Texts and Cultures: Nordic Dialogues* (2018).

Perspectives on the Non-Human in Literature and Culture

Series Editor: Karen Raber, *University of Mississippi, USA*

Literary and cultural criticism has ventured into a brave new world in recent decades: posthumanism, ecocriticism, critical animal studies, the new materialisms, the new vitalism, and other related approaches have transformed the critical environment, reinvigorating our encounters with familiar texts, and inviting us to take note of new or neglected ones. A vast array of non-human creatures, things, and forces are now emerging as important agents in their own right. Inspired by human concern for an ailing planet, ecocriticism has grappled with the question of how important works of art can be to the preservation of something we have traditionally called "nature." Yet literature's capacity to take us on unexpected journeys through the networks of affiliation and affinity we share with the earth on which we dwell—and without which we die—and to confront us with the drama of our common struggle to survive and thrive has not diminished in the face of what Lyn White Jr. called "our ecological crisis." From animals to androids, non-human creatures and objects populate critical analyses in increasingly complex ways, complicating our conception of the cosmos by dethroning the individual subject and dismantling the comfortable categories through which we have interpreted our existence. Until now, however, the elements that compose this wave of scholarship on non-human entities have had limited places to gather to be nurtured as a collective project. "Perspectives on the Non-Human in Literature and Culture" provides that local habitation. In this series, readers will find creatures of all descriptions, as well as every other form of biological life; they will also meet the non-biological, the microscopic, the ethereal, the intangible. It is our goal for the series to provide an encounter zone where all forms of human engagement with the non-human in all periods and national literatures can be explored, and where the discoveries that result can speak to one another, as well as to scholars and students.

Plants in Children's and Young Adult Literature
Edited by Melanie Duckworth and Lykke Guanio-Uluru

Animal Remains
Edited by Sarah Bezan and Robert McKay

The Landscapes of W. H. Auden's Interwar Poetry
Roots and Routes
Ladislav Vít

For more information about this series, please visit: www.routledge.com/Perspectives-on-the-Non-Human-in-Literature-and-Culture/book-series/PNHLC

Plants in Children's and Young Adult Literature

Edited by Melanie Duckworth and Lykke Guanio-Uluru

Routledge
Taylor & Francis Group
NEW YORK AND LONDON

First published 2022
by Routledge
605 Third Avenue, New York, NY 10158

and by Routledge
2 Park Square, Milton Park, Abingdon, Oxon OX14 4RN

Routledge is an imprint of the Taylor & Francis Group, an informa business

Library of Congress Cataloging-in-Publication Data
Names: Duckworth, Melanie, editor. | Guanio-Uluru, Lykke, editor.
Title: Plants in children's and young adult literature / edited by Melanie Duckworth and Lykke Guanio-Uluru.
Description: New York, NY : Routledge, 2021. |
Series: Perspectives on the non-human in literature and culture | Includes bibliographical references and index.
Identifiers: LCCN 2021021585 | ISBN 9781032066349 (hardback) | ISBN 9781032122458 (paperback) | ISBN 9781032066356 (ebook)
Subjects: LCSH: Plants in literature. | Children's literature--History and criticism. | Young adult literature--History and criticism. | LCGFT: Literary criticism.
Classification: LCC PN1009.5.P58 P53 2021 | DDC 809/.93364083--dc23
LC record available at https://lccn.loc.gov/2021021585

ISBN: 978-1-032-06634-9 (hbk)
ISBN: 978-1-032-12245-8 (pbk)
ISBN: 978-1-032-06635-6 (ebk)

DOI: 10.4324/9781032066356

Typeset in Sabon
by Taylor & Francis Books

This volume is a tribute to all the species of plants with which we share our lives, every day and throughout our lives. We acknowledge all the ways in which you sustain us and other forms of life on this planet, as well as our symbiotic relationship through breath: what you breathe out, we breathe in. What we breathe out, you breathe in. Encountering plants on the pages of our literatures, the material vehicle of which are also plants, helps us reflect on and appreciate this interrelationship.

Contents

Figures

Contributors

Katri Aholainen is a doctoral student of Finnish literature at the University of Turku, Finland. She studies Finnish author Yrjö Kokko's works by focusing on non-human agencies in processes of writing and photographing. Her theoretical frameworks include posthumanism and material feminism, especially Karen Barad's agential realism.

Francesca Arnavas completed her PhD at the University of York and is now a research fellow at the University of Tartu. Her research interests are Victorian literature, cognitive narratology, and fairy tale studies. Her first book is *Lewis Carroll's "Alice" and Cognitive Narratology: Author, Reader and Characters* (De Gruyter, 2021).

Justyna Deszcz-Tryhubczak is Associate Professor of Literature at the Institute of English Studies, University of Wroclaw, Poland. She published *Yes to Solidarity, No to Oppression: Radical Fantasy Fiction and Its Young Readers* (2016) and co-edited *Intergenerational Solidarity in Children's Literature and Film* (2021).

Terri Doughty teaches in the English Department at Vancouver Island University, Canada. She is the editor of *Selections from the Girl's Own Paper 1880–1907* and co-editor of *Knowing Their Place? Identity and Space in Children's Literature*. She has published on girls' empire adventures, young adult fantasy, and the work of Brian Selznick.

Melanie Duckworth is Associate Professor of English Literature at Østfold University College, Norway, where she teaches British, postcolonial, and children's literature. Her research interests include Australian literature, plant studies, children's literature, and ecocriticism and she has published on Australian historical children's fiction, Australian literature, ecofeminism, and contemporary poetry.

Hallvard J. Fossheim is Professor of Philosophy at the University of Bergen. He has published widely on Plato and Aristotle; among his recent pieces are "*To kalon* and the Experience of Art" in *The Poetics in Its*

Aristotelian Context (Routledge, 2020), and "The Number of Rulers in Plato's *Statesman*", *Polis* (2020).

Nina Goga is Professor of Children's Literature at Western Norway University of Applied Sciences. Her latest co-edited books are *Verbal and Visual Strategies in Nonfiction Picturebooks: Theoretical and Analytical Approaches* (2021) and *Ecocritical Perspectives on Children's Texts and Cultures: Nordic Dialogues* (2018).

Lykke Guanio-Uluru is Professor of Literature at Western Norway University and researches literature and ethics, particularly plant studies, ecocriticism, fantasy, and game studies. She is the author of *Ethics and Form in Fantasy Literature* (2015) and multiple research articles, and co-editor of *Ecocritical Perspectives on Children's Texts and Cultures: Nordic Dialogues* (2018).

Andrea Casals Hill has a PhD in Literature from Universidad Católica de Chile, where she is conducting a three-year research project on YA dystopian narratives. She is co-editor of the academic journal *English Studies in Latin America* and recently published *Futuro esplendor: ecocrítica desde Chile* (2019), co-authored with Pablo Chiuminatto.

Anja Höing is a lecturer at Osnabrück University, Germany. Her main research interests lie in representations of childhood and the natural world in literature. Recent publications include contributions to the edited collections *Romanticism and the Cultures of Infancy* (Palgrave Macmillan, 2020) and *Gothic Animals* (Palgrave Macmillan, 2020).

Alida Mayne-Nicholls (MA, PhD) is a professor of literary criticism, poetry, and children's representations in films and literature at the Liberal Arts programme at Universidad Adolfo Ibáñez, Santiago, Chile. Her main areas of research are childhood studies and feminism. She has published articles focused on poetry, women's studies, and childhood studies.

Ivana Mijić Nemet received her PhD in 2021 at the University of Novi Sad, Faculty of Philosophy, with the thesis "Poetic features of the contemporary Serbian fantasy novel for children". Since 2012 she has been secretary of the journal for children's literature *Detinjstvo*. Her main research interest is children's and fantasy literature and she publishes essays, studies, and literary criticism.

Mónika Rusvai is a fantasy author and a PhD student at the University of Szeged, Hungary. Her MA, drawing on monster theory, discussed the cultural significance of various European dragons. She now uses critical plant studies and fantasy theory to research enchanted forests, currently focusing on Robert Holdstock's Mythago novels.

Jose Monfred C. Sy teaches in the Department of Filipino and Philippine Literature at the University of the Philippines Diliman. His research appears in *Literary Cultures across the World* (2018) and *Ecologies in*

Southeast Asian Literatures: Histories, Myths and Societies (2019). He is a volunteer teacher for displaced Indigenous youth through the Save Our Schools Network. He has also published children's books.

Tijana Tropin (PhD) is a research associate at the Institute for Literature and Arts in Belgrade, Serbia. Her interests are in comparative literature and literary theory, translation studies, children's literature, and fantasy literature. She is the author of *Motiv Arkadije u dečjoj književnosti* [The Arcadian Motif in Children's Literature and co-editor of *Časopisi za decu: Jugoslovensko nasleđe (1918–1991)* [Children's Periodicals: Yugoslav Heritage (1918–1991)].

Charlotte van Bergen has an MA in Children's Literature from Tilburg University and completed a research master's in Historical, Literary, and Cultural Studies at Radboud University, the Netherlands. She was an intern at the Centre for Research on Children's and Young Adult Literature at the University of Wrocław, Poland. She works for the Dutch Research Council and is continuing her research on the role and representation of fathers in children's literature from the past century.

Preface

All along, it has felt as if this book wanted to exist. Its seed was planted during a chance meeting in the cafeteria of the Oslo University Library many years ago, although we did not know this at the time. There for the same job interview, we happened to briefly connect over coffee. It was very much later, as Melanie sat at her office in the small Norwegian town of Halden, looking out over the oak trees beyond the car park and reading Lykke's essay "Plant-Human Hybridity in the Story World of Kubbe" in *Ecocritical Perspectives on Children's Texts and Cultures: Nordic Dialogues* (2018), that the realization occurred: "I've met her!" Pondering the possibility of putting together a research volume purely about trees in children's literature, Melanie decided to reach out to Lykke with her request. As it turned out, Lykke had already been entertaining similar ideas. "Why not a volume about *plants*, not just trees, and how about including young adult literature?" she suggested, and thus the seed begun to sprout.

The overwhelming number of scholars who responded to our call for papers demonstrated that phyto-awareness was growing in more minds than ours.

It has been a genuine joy to work with this volume and its international contributors. Plants migrate widely, and we are happy that we were able to include tales of plant encounters from so many different places. We are grateful for the enthusiasm and dedication of our series editor, Karen Raber, and our commissioning editor, Jennifer Abbott, who both embraced our proposal from the start, as well as to the anonymous peer reviewers, whose comments helped strengthen each contribution to the volume. The encouraging reports of the reviewers of the full manuscript were surely heartening—and so the seed has reached maturity. The first international volume on plants in children and young adult literature is now in your hands. Who knows where its seeds will take root and grow?

Melanie Duckworth
Lykke Guanio-Uluru

Introduction

Plants in Children's and Young Adult Literature

Lykke Guanio-Uluru and Melanie Duckworth

In *Anne of Green Gables*, Lucy Maud Montgomery shows Anne imaginatively engaging with the idea of plant communication: "Wouldn't it be nice if roses could talk? I'm sure they would tell us such lovely things" (1908, p. 45). Anne's romantic musings appear in stark contrast to the imaginings of an Old English poet, who centuries earlier invokes the plaintive voice of the tree fashioned into the cross of Christ: "They drove dark nails into me; the dints of those wounds can still be seen / Open marks of malice; but I did not dare maul any of them in return" ("The Dream of the Rood", 8th–10th century, unknown author). Framed from the point of view of the tree, "The Dream of the Rood" echoes a sense of vegetal hardship akin to that voiced by the Yggdrasil Ash in Old Norse mythology, and later by the morose Old Forest in J. R. Tolkien's *The Lord of the Rings*.

These few brief examples testify to a long line of literary thought engaging with the notion of plants as communicators and fellow beings or "kin", affecting and being affected by their interrelationship with humanity. Curiosity about the lifeworld of plants is currently echoed and affirmed by the rapidly expanding field of critical plant studies. In parallel with recent biological research, which draws on both historical and more recent studies to demonstrate the variety and complexity of the adaptive behaviour of plants (Mancuso & Viola, 2016; Pollan, 2013; Gagliano, 2018), the representation of plants in science, arts, and literature has come under scrutiny.

One probable reason for the current and intensified interest in the cultural and biological role of plants is the growing awareness of the detrimental effects wrought by human activity on the Earth's natural systems, which includes its effects on marine flora and fauna. While some term this historical moment the Anthropocene,[1] the knowledge that "plants dominate every terrestrial environment, composing ninety-nine percent of the biomass on Earth" (Mancuso & Viola, 2016, p. xii) should serve to instil some humility into anthropocentric planetary thinking, while we should still take appropriate responsibility for the effects of human activity upon other species.

The upsurge of interest in plants, plant sensing, and plant rights may also be regarded as the cumulative effects of ecocriticism (with Carson, 1962; Næss, 1989; Buell, 1995; 2005, and Glotfelty & Fromm, 1996 as early and

DOI: 10.4324/9781032066356-1

influential works), animal activism (Singer, 2002; Derrida, 2002; Haraway, 2008), and critical and feminist posthumanism (Haraway, 1991; Hayles, 1999; Wolfe, 2010; Braidotti, 2013). These -isms have all gained momentum as persistent critiques of the modern capitalist exploitation of both the environment and other species by calling for a revision of the anthropocentric world view characteristic of current Western culture. While there are now a number of works analysing children's and young adult (YA) literature from an ecocritical perspective (see for instance Dobrin & Kidd, 2004; Curry, 2013; Goga et al., 2018; Ryan, 2018), this is the first volume within the field that focuses exclusively on the representation of plants in literature for children and young adults, although some, earlier but single, studies exist (Hines, 2004; Jaques, 2015; Kokkola, 2016; 2017; Goga, 2017; 2020; Guanio-Uluru, 2018; 2019). Given the current and widespread interest in plant representation, we believe that this collection is a timely contribution to the fields of children's and YA literary research that will also enrich this Perspectives on the Non-Human in Literature and Culture series.

Critical Plant Studies

An early proponent of vegetal life was Christopher D. Stone, authoring *Should Trees have Moral Standing*, first published in 1972. More lately, plant-thinking has influentially been taken up by Michael Marder, publishing *Plant-Thinking: A Philosophy of Vegetal Life* and editing the anthology *Critical Plant Studies: Philosophy, Religion and Culture* in 2013, publishing *The Philosopher's Plant: An Intellectual Herbarium* in 2015, and *Grafts: Writings on Plants* in 2016. In 2016 he also co-published *Through Vegetal Being: Two Philosophical Perspectives* with Luce Irigaray. Pre-dating Marder was the examination by Matthew Hall in 2011 of human attitudes to plants across a range of global philosophical systems, including indigenous and pagan systems of thought, which have long retained more respectful alliances with the plant kingdom than has Western science.

Hall finds one of the "roots of disregard" for plants in Greek philosophy, and most notably in Plato's zoocentric description of plants as created not for their own sake but as a human and animal utility, as well as in Aristotle's "bracketing of plants into a lower class of being" (Hall, 2011, p. 22) through his influential theory of the tripartite soul; a reading of Aristotle with which Hallvard Fossheim takes issue in this volume's third chapter. Discussing Christianity, Hall also calls attention to the fact that in Genesis, the concept of "living creatures" include sea and land animals, but not plants (2011, p. 57). Thus, when Noah is instructed to collect two of every sort of "living thing" on his ark to keep them alive during the great flood, plants are excluded. As Hall remarks: "A more appropriate ecological escape plan would have included plants on the ark" (2011, p. 59). It is hard to disagree, given the essential part plants play in local and global ecosystems, and their importance to Earth's biodiversity (Schramski et al., 2015).

Revisiting Aristotle by way of a discussion of Michel Foucault's concept of biopower, Jeffrey T. Nealon demonstrates that the lack of interest in discussing plants as a form of life, or bios, has persisted in Western philosophy: "[I]t became clear to me that the plant, rather than the animal, functions as the life forgotten and abjected within a dominant regime of humanist biopower", Nealon says (2016, p. x). The zoocentrism evident in both early Greek and Christian systems of thought seems to have manifested in a condition that J. H. Wandersee and E. E. Schuster (1999) have labelled "plant blindness." This condition is marked by a tendency towards not seeing, and therefore not appreciating, plants. Wandersee and R. E. Clary (2006) have later defined plant blindness more specifically as the inability to see or recognize the plants in one's environment, frequently combined with an inability to recognize the unique biological features of plants and to appreciate their importance in the biosphere and in human affairs. Wandersee and Clary regard this as a universally human, biological "default condition" (2006, p. 2) and argue that it has developed in the course of evolution, since plants generally tend to pose less of a treat to human survival than do wild animals that therefore more easily grab our attention. Hall rejects this proposition, holding that zoocentrism "is largely a cultural-philosophical attitude" (2011, p. 6). He argues that while both Eastern and Western philosophical systems fail to acknowledge plants as "persons" and thus as worthy of respect, some versions of Hinduism, along with several indigenous traditions and some pagan cultures, conceive of plants as kin.

Within the field of ecocriticism, research on kinship is furthered in particular by Donna Haraway, not least in her latest book *Staying with the Trouble, Making Kin in the Chthulucene* (2016). Calling for the breaking down of established dichotomies in her "Cyborg manifesto" (1991), Haraway has persistently argued for an inclusive attitude in interspecies relations. However, the more explicit inclusion of plants into her interspecies thinking is recent. While her work on pets as "companion species" in *When Species Meet* (2008) has no index entry for plants, her latest work renames the humanities the "humusities" (p. 32) and contains several index words indicative of the influence of critical plants studies on her thought (such as phytolinguistics, plantation system, reforestation, plants, plant sympoeisis, and compost). Thinking in terms of string figures, Haraway concedes to being "unhappy with posthumanism" (p. 32) and envisions a "niche space" for all critters, a "Terrrapolis" that is "not 'posthuman' but com-post'" (p. 11). A similarly influential thinker of the posthuman, N. Katherine Hayles equally looks to plant theory in her latest publication, *Unthought, The Power of the Cognitive Nonconscious* (2017), in which she includes a discussion of plant signalling and recent claims of plant intelligence in her querying of nonconscious cognition (pp. 16–20).

With established tropes of anthropomorphism, wonder, and connection with the natural world, children's literature is another medium through which the notion of kinship with plants has long been approached. In

Western literatures, the legacy of animism and paganist views on plants as sentient beings is particularly visible in the folkloric genre of fairy tales. Fantasy literature, which builds upon this fairy tale tradition, as well as on creation myths from multiple cultures, has carried on this legacy into the present day. Tolkien's walking and talking tree guardians, the Ents, are modern icons of plant agency, inspired both by Tolkien's own deep love of trees and by motifs of talking trees from Old Norse mythology (Sturluson, 2005) and from Anglo-Saxon poetry, like the aforementioned "Dream of the Rood" (see Guanio-Uluru, 2015, and Ryan, 2016). Modern fantasy literature may be regarded as the stepchild of the Enlightenment, in the sense that pagan, animist, and religious world views migrated to this genre in the wake of the Enlightenment emphasis on rationality and sense as governing impulses.

Generally, trees seem to hold a privileged place in the plant kingdom in the beliefs and imaginations of many peoples. In *The Golden Bough*, James Frazer notes that:

> In the religious history of the Aryan race in Europe the worship of trees has played an important part. Nothing could be more natural. For at the dawn of history Europe was covered with immense primaeval forests, in which the scattered clearings must have appeared like islets in an ocean of green.
>
> (p. 109)

Sarah Maitland argues that the genre of the fairy tale is deeply entwined with the forest as a particular form of landscape: "[F]orests and fairy stories are like trees and their mycorriza ... The Fairy stories teach us how to see the forests, and how to love them too" (2012, p. 21). Given this rich and storied vein of folklore, and the folkloric legacy of fantasy literature, it is no surprise that this collection contains several chapters analysing fairy tales and fantasy stories. Part II in this volume "Plants in Folklore and Fantasy" picks up on this heritage and examines the representation of plants in these traditional genres. The primary texts analysed in Part III: "Arboreal Embraces", may also be read as continuations of indigenous, pagan and folkloric traditions, where plants, and most often trees, are celebrated as kin.

A fascination with plants is of course an integral part of the field of botany. A digging for roots demonstrates that thoughts on plant cognition are not new within the sciences either. Erasmus Darwin (1796) argued that plants are akin to animals and have sensation and movement. His more famous grandson, Charles Darwin, was convinced that plants' roots function akin to a brain (1880). More recent research has confirmed Erasmus Darwin's argument that plants have a number of highly developed senses, among them an acute sensitivity to light, and to different light qualities (Mancuso & Viola, 2016, p. 47).

Plant behaviours, such as the tendency to grow towards the light, and the strategy of increasing the speed of growth to outgrow neighbouring rivals, further demonstrate that plants plan and use resources to bring about future ends (Mancuso & Viola, 2016, p. 49). Stefano Mancuso and Alessandra Viola discuss how the plant cycle of shedding leaves, which contain most of the plant's light receptors or "eyes", are a form of hibernation or rest: plants "close their eyes" and "sleep" during the winter, and reopen their "eyes" in spring, growing new foliage (2016, p. 52). Plants also communicate, through pheromones—for instance, researchers have found that across species, plants release the substance of methyl jasmonate when they experience stressful conditions, also to warn neighbouring plants of imminent dangers (Mancuso & Viola, 2016, pp. 55–56). The first part of this volume, "Botanical Fascinations" specifically explores children's literary texts that display a concern with imparting botanical knowledge through metaphor as well as through verbal and visual representation.

While the common stereotypical view of plants is that they are sessile and "passive", several plants hunt to feed, such as the Fly Trap Sensitive (*Dionaea muscipula*) that traps flies. This fact was known to Western science already in the 19th century, and was well documented by Charles Darwin in *Insectivorous Plants*, published in 1875. So far, more than 600 carnivorous plant species are documented. They use a number of different hunting techniques and devices, including underground traps, to catch a variety of insects and even small animals, such as lizards and mice, which they digest using enzymes (Mancuso & Viola, 2016, pp. 64–65). Such species of carnivorous plants, as documented by Darwin, conceivably gave rise to the literary trope of the monster plant, explored in such tales as John Wyndham's *The Day of the Triffids* (1951) and discussed in this volume's Chapter 6.

Nealon notes how the metaphorical model of the tree has been central to biology (evolutionary models), philosophy (truth trees), and linguistics (trees of language development) (2016, pp. 87–88). This mapping of knowledge onto the model of the tree was famously criticized by Gilles Deleuze and Félix Guattari in *A Thousand Plateaus* (1987), in which they reject "arborescent" structures of knowledge as hierarchic and replace them with their influential metaphor of the rhizome. While global forests are now threatened, the metaphor of the rhizome has proven as resistant and invasive as the types of plant from which the metaphor is derived. This replacement in the organization of our thinking with one type of plant for another aptly displays our tendency to model our thought along vegetal patterns and calls to mind Yuval Nova Harari's argument that not only have we domesticated plants, but certain species have also "domesticated" us: "Worldwide, wheat covers about 2.25 million square kilometres of the globe's surface, almost ten times the size of Britain. How did this grass turn from insignificant to ubiquitous? Wheat did it by manipulating Homo Sapiens to its advantage" (Harari, 2011, p. 90).[2]

In this vein, more recent scientific descriptions of "agentic" plants reconfigure plants as active beings rather than as merely food, props, and the background for animal lives. The word "agency", that is, the capacity to act, is central in reimagining plants as beings valid in their own right. According to Australian eco-philosopher Val Plumwood, to view "earth others as fellow agents and narrative subjects is crucial for all ethical, collaborative, communicative and mutualistic projects" (2002, p. 175). As well as being a significant concept in plant studies, the idea of "agency" is a central concern of material ecocriticism, which envisions a world "vibrant" with the agency of things as well as beings (Iovino & Oppermann, 2014; Barad, 2007; Bennett, 2010). It is also an important concept in children's literature and childhood studies, where the real and imagined agencies of children are a highly contested topic (Gubar, 2013, 2016; Castro & Clark, 2019). The final section of this book, "Plant Agency and Activism", specifically addresses representations of agentic plants, but the intersecting agencies of plants, children, and other beings are woven through all contributions to the volume.

Ecocritical and Plant-Centred Approaches to Children's and YA Literature

Monica Gagliano, John C. Ryan, and Patrícia Vieira note that in the Western poetry and prose tradition, plants tend to be represented "as part of the landscape or as the backdrop for human and, on occasion, animal dramas" and to become "the correlatives of human emotions, eliciting feelings of pleasure and displeasure, triggering memories, and reflecting human states of mind" (Gagliano et al., 2017, p. x), so that literary discourse predominantly renders the lives of plants as "symbolic or figurative, as organic referents for animal meaning" (p. xi). On the other hand, Ryan has investigated contemporary poetry as a field rich with possibilities for engaging imaginatively with plants (2018). Despite this, plants are underrepresented in literature, as in scientific studies. As Randy Laist notes: "The scarcity of plant-life in the cultural canon of the contemporary West is particularly striking when contrasted against the ubiquity of stories that feature animals as characters, subjects, and symbols" (2013, p. 11). The prevalence of stories about animal characters is a trait much commented upon in children's literature research.

Ecocritical approaches to children's literature first came to prominence with the publication of Sidney I. Dobrin and Kenneth B. Kidd's edited collection *Wild Things: Children's Culture and Ecocriticism* in 2004. Before this, there had been a few special issues of journals dedicated to the theme (the spring 1995 issue of the *American Nature Writing Newsletter*; the *Children's Literature Association Quarterly 19.4 (1995)*, with a special section on "Ecology and the Child" edited by Betty Greenway; and the 1995 special issue of *The Lion and the Unicorn*, "Green Worlds: Nature and Ecology", edited by Suzanne Rahn). Two out of the 16 essays in *Wild Things* focused specifically on plants (Hines, 2004; Henderson, Kennedy &

Chamberlin, 2004). Richard Li and Ryan comment on the 1995 special issue of *The Lion and the Unicorn*: "Although the issue's first cluster of articles is entitled "The Forest", the content of the issue as a whole remains conspicuously zoocentric" (2018, p. 232).

While ecocritical approaches to children's literature have been gaining traction in recent years (Bradford et al., 2008; Massey & Bradford, 2011; Cutter-Mackenzie et al. 2011; Curry, 2013, Buell, 2014), in keeping with wider trends, more focus has certainly been given to the representation of animals than that of plants. As Greta Gaard puts it: "Of necessity, given the predominance of animal fables in children's literature, this branch of ecocriticism has developed an emphasis interrogating the relationship of culture and nature through the relationships of children and animals" (2008, p. 325). Two notable exceptions to the animal-centric focus of ecocritical approaches to children's literature have been the sustained attention given to two North American picturebooks: Theodor Seuss Geisel's *The Lorax* (1971/2012) (Marshall, 1996; Graulich, 1996; Ross, 1996; Henderson, Kennedy and Chamberlin, 2004; Gaard, 2008; Bergthaller, 2006) and Shel Silverstein's *The Giving Tree* (1964) (Jaques, 2015; Gaard, 2008; Miller, 2012). *The Lorax* is an environmental fable of ecological destruction, featuring the Lorax who claims to "speak for the trees", and the "Once-ler" who chops them all down to convert them into profit. *The Giving Tree* features a female coded tree cheerfully sacrificing herself to the needs of the boy—and then the man—she loves. *The Lorax* ends with a grain of hope in the form of a seed, and *The Giving Tree* concludes with the thoroughly discomforting assertion that the stump of the tree is "happy" in her new role as a seat for her beloved. The destruction of trees in these two books can be read as standing in for the environment as a whole. While the Lorax is a story of failing to save trees, most discussions about the book have centred around its environmental rhetoric, and the dubious device of passing on environmental responsibility to the next generation. The prominence of these two texts testifies to the North American location of the development of ecocriticism—as Bergthaller points out, *The Lorax* was until recently little-known outside of the United States (2006, p. 160). The books are mentioned as touchstones by most Anglo-American forays into plants studies in children's literature to date (Jaques, 2015, p. 116, pp. 126–131; Li & Ryan, 2018, p. 233, p. 234; Kokkola, 2017, p. 276).

Recently, a few critics have made the case for the integration of plant studies with children's literature studies, most notably Lydia Kokkola, Guanio-Uluru, Ryan, and Zoe Jaques. These critics also demonstrate the need to go beyond North American texts to explore the representations of plants in a variety of international local contexts. Kokkola offers a brief overview of potential connections between "Critical Plant Studies and Children's Literature" in the *Edinburgh Companion to Children's Literature* (2017). She points out that while active and agentive plants do figure widely in children's fiction (though not as widely as animals do), they still tend to be connected to human agendas. Jaques's *Children's Literature and the Posthuman: Animal, Environment, Cyborg* (2015) contains a chapter on trees. Like Kokkola, she notes that they are most often represented as being in service to humans but argues: "that service is replete with direct opposition and resistance. This plant rebellion, although often contained, generates unease about the rights of humanity to "use" trees and anxiety about a human superiority with notably selfish ends" (2015, p. 117).

In "Sowing Seeds: Phytocriticism and the Botanical Dimensions of Indonesian literature for Children and Young Adults", (2018) Li and Ryan discuss the representation of plants in Indonesian children's literature. Their survey, limited to Indonesian texts translated into English or published as dual language texts, aims to address a "pronounced absence of scholarly attention to the environmental dimensions of Asian and, most glaringly, Southeast Asian works" (p. 231). *Beasts at Bedtime: Revealing the Environmental Wisdom in Children's Literature* (2018), written for a general audience by zoologist Liam Heneghan, privileges animals in its title and within its pages, but includes considerations of plants in some of its chapters. The plants discussed in the book, besides the compulsory *Lorax* and *The Giving Tree*, include gardens in the stories of Hans Christian Andersen, a "plant motif" running through the Brothers Grimm's story of Rapunzel, and the significance of a botanical education in the training of wizards in fantasy literature such as Ursula le Guin's *Earthsea* trilogy and J. K. Rowling's *Harry Potter* series. Heneghan also chooses to end the book with a reference to an alternative North American tree-conservation classic, Mildred Taylor's *Song of the Trees* (1975/2016), in which "the Logan family intervenes to prevent their white neighbour, Mr Anderson, from cutting down trees on their property" (p. 276), thus drawing attention to the, often overlooked, significance of the relationship between black children and plants.

Ecocritical Perspectives on Children's Texts and Cultures: Nordic Dialogues by Nina Goga, Guanio-Uluru, Bjørg Oddrun Hallås, and Aslaug Nyrnes, focuses on environmental concerns in contemporary Nordic children's literature. Amid other essays focusing on landscapes, aesthetics, and animals, it includes two essays under the heading "vegetal". Guanio-Uluru's chapter, "Plant-Human Hybridity in the Story World of Kubbe", draws on posthumanism and critical plant studies to problematize the character of Kubbe, an anthropomorphized log who features in a picturebook series by Norwegian author and illustrator Åshild Kanstad Johnsen. Anna Karlskov Skyggebjerg's chapter "Poetic Constructions of Nature: The Forest in Recent Visual Poetry for Children", examines the Danish author and illustrator Karen Filskov's *Skoven fra oven* (The Forest From Above) as a playful text which uses visual poetry to represent contemporary forests while also drawing on tropes from fairy tales. Additionally, Goga (2017) has analysed variations in the visual representation of Pinocchio from a posthuman perspective, drawing on Jaques and focusing on the representation of

Pinocchio's formal shifts from plant (wood) to animal (donkey) to human. Further extending the remit of the intersection between plant studies and YA literature in "Imagining Climate Change: The Representation of Plants in Three Nordic Climate Fictions for Young Adults" (2019), Guanio-Uluru has tallied the number of and analysed the ways in which plants are represented in three Nordic climate novels for young adults, showing that, despite the importance of plants to the global climate, the degree to which plants feature in these fictions varies considerably.

The confident assertion of the significance of animals over plants in children's literature may itself be a case of "plant blindness". Plants are frequently relegated to the background, or environment, of literature, but on the other hand, they are also used to demarcate spaces for childhood exploration and discovery, in the Romantic tradition. Enid Blyton's *The Magic Faraway Tree* (1943/2016), for example, is not only home to a number of magical creatures but reaches to the clouds, has magical properties, and acts as a ladder to fantastic lands. The eponymous "wild things" of Maurice Sendak's *Where the Wild Things Are* (1964/2000), refer to the animalistic monsters Max meets at the end of his journey. However, the words could just as well indicate the trees and vines that grow in his room, the jungle that awaits him, and the colourful leaves that crown the end-pages of the book. Max's journey could not have begun without the strange and accelerated agency of the forest which "grew / and grew / until the ceiling hung with vines / and the walls became the wall all around" (Sendak, 1964/2000, unpaginated). It is plants that transform "walls" into "the world all around", and the tone of these words invites a more capacious interpretation of "world" than simply "background".

The Organization of this Volume

In this volume at least, plants are at the centre, rather than at the background, of scholarly attention. The overarching question of this anthology is: How are plants represented in children's and young adult literatures? This main question gives rise to more specific questions: Are plants treated primarily as parts of established literary landscape topoi, such as the garden, the forest, and the wilderness, notable in literary classics such as *The Secret Garden* (2012) by Frances Hodgson Burnett, Elsa Beskow's *Children of the Forest* (1910/1986), or *Where the Wild Things Are*, or do they emerge predominantly as the supporting structures of the literary action, as in Andy Griffiths and Terry Denton's *Treehouse* books (2011–), or as portals to fields of learning, as in Mary Pope Osborne's *The Magic Treehouse* series (1992–)? Do the verbal text and the images of illustrated children's stories and picturebooks tell the same, or divergent, stories about plant life and plant agency? Are there plant tropes common to particular genres, such as fairy tales and fantasy literature? Does plant representation in such literatures contradict or sanction the anthropocentric paradigm inherent in most Western literatures and cultures? Are there local, chronological, and generic specificities in the ways that plants are represented, or is the main difference rather to be found in the analytical perspectives with which academics approach these texts?

The present anthology has aimed to include researchers and primary texts from a variety of global contexts. The collection includes contributions from scholars in Europe, South America, North America, and the Philippines. The book is arranged thematically into four parts, based on different ways of thinking about or apprehending plants. Most sections have a broad geographical remit, and contain texts aimed at both older and younger readers. The first part, "Botanical Fascinations" comprises chapters which engage with plant botany and the philosophical distinction between plants and other species. In Chapter 1, "A Relational Poetics of Plant–Human Interactions: Contrasting the Picture Books of Cicely Mary Barker and Elsa Beskow", Terri Doughty contrasts the "Flower Fairy" picturebooks of English artist Cicely Mary Barker and selected picturebooks by the Swedish artist Elsa Beskow. Doughty finds that Beskow, unlike Barker, represents plants living within complex communities and models human children interacting in non-exploitive ways with the flora, modelling a cognitive shift from the Anthropocene to what Haraway terms the Chthulucene. In Chapter 2, "Stamens and Pistils in the Same Flower: Queer Performativity of Plants in Finnish Fairy Tale 'Pessi ja Illusia'", Katri Aholainen scrutinizes how the biology of plants is entwined with the fairy tale tradition, where plant characters who have "stamens and pistils in the same flower" convey a different message to the text's heteronormative animal couples. Concerning itself with philosophical distinctions between plants and humans, Chapter 3, "Aristotle on Plants: Life, Communion, and Wonder" by Hallvard Fossheim, takes a side-step from directly examining children's and young adult literature to address the repeated accusation within critical plant studies that Aristotle philosophically devalued plants. In a close reading of Aristotle, Fossheim argues that the philosopher should rather be regarded as advocating wonder towards and communion with the vegetal world, since to Aristotle humans are all, at least partly, plants. With a basis in Aristotle's use of analogy, Fossheim further notes that literary representations such as fantasy and fairy tales may provide a different kind of insight into the life world of plants than that afforded by empirical studies or philosophy—a point duly demonstrated and discussed in the second part of the book, "Plants in Folklore and Fantasy", which focuses on the roles of plants in these significant genres.

In Chapter 4, "Come into the Garden, Alice: Rude Flowers, Dream-Rushes, Aphasic Woods, and Other Plants in Lewis Carroll's Nonsense Worlds", Francesca Arnavas enquires into the complex roles and narrative functions of plants in Carroll's *Alice in Wonderland* and *Through the Looking-Glass*, with reference to the motif of the enchanted forest in the children's literature tradition and the Victorian obsession with the language of flowers. Chapter 5, "Fern Blossom and Lilibala: Magical Plants in Serbian

Children's fantasy", by Tijana Tropin and Ivana Mijić Nemet, establishes an overview and analysis of fantastic and magical plants and of their heterogeneous origins and functions in contemporary Serbian fantasy for children, which borrows from Serbian folklore. In Chapter 6, "Vegetal Magic: Agnieszka's Journey to the Understanding of the Vegetal Other in Naomi Novik's *Uprooted*", Mónika Rusvai analyses Novik's construction of vegetal magic as analogous to Jane Bennett's (2010) concept of vital materiality.

The longstanding human fascination with trees and forests remains a feature of modern children's literature. Part III, "Arboreal Embraces", which is the most voluminous section, focuses on human relationships with trees. In Chapter 7, "Arboreal and Maternal Desires: Trees and Mothers in recent Australian Middle-Grade Fiction", Melanie Duckworth examines the ways in which the desires of mothers and trees are juxtaposed, merged, and contrasted in three recent Australian middle-grade novels, presenting alternative versions of the trope of the "mother tree" to Silverstein's sacrificial *The Giving Tree* (1964/2014). In Chapter 8, "Arboreal Entanglements: Childrenforest and Deforestation in Ecopoetry by Children", Justyna Deszcz-Tryhubczak and Charlotte van Bergen engage with the concept of childrenforest through and with ecopoems about deforestation written in 2018 by pupils of John Clare Primary School in the village of Helpston, Northamptonshire, England. Looking at these poems as participating in an assemblage of human and non-human entities, they rely on critical plant studies and botanical criticism to explore the childrenforest co-constituted by these texts. In Chapter 9, "'I felt like a tree lost in a storm'—The process of entangled knowing, becoming, and doing in Beatrice Alemagna's picturebook *Un grande giorno di niente*" (2016), Nina Goga focuses on how the child protagonist entangles with trees, by mapping out how various forms of nature/culture binaries are depicted and blurred in Alemagna's picturebook, and argues that the boy protagonist's various interactions with trees in both the text and the illustrations challenge his previous conception of outdoor life, and of his relationship or kinship with the woods. In Chapter 10, "From Chamomiles to Oaks: Agency and Cultivation of Self-Awareness", Alida Mayne-Nicholls and Andrea Casals Hill discuss two Spanish texts for children that feature the agency of plants. The chapter explores the rhetorical and visual devices deployed by the authors to represent both chamomiles and oaks as active communicative subjects, participating with children in a nonlinear, informal process of cultivating self-awareness.

Thus, Mayne-Nicholls' and Casals Hill's chapter provides a bridge to the volume's fourth and last part that discusses "Plant Agency and Activism". In Chapter 11, "Vegetal Individuals and Plant Agency in Twenty-First Century Children's Literature", Anja Höing examines three recently published middle-grade novels featuring vegetative individuals: Kathleen Van Cleve's *Drizzle* (2010), Frances Hardinge's *The Lie Tree* (2015), and Sophie Anderson's *The Girl Who Speaks Bear* (2019). The chapter explores the question of how far such fictional representations of plant-child interaction may suggest new ways of making sense of human–plant ties. The volume's

Chapter 12, "Vegetable Violence: The Agency, Personhood, and Rhetorical Role of Vegetables in Andy Griffiths and Terry Denton's *The 52-Storey Treehouse*" by Lykke Guanio-Uluru, discusses the fourth book in this bestselling Australian illustrated book series. In a series criticized for its subversive and scatological humour, *The 52-Storey Treehouse* features multiple episodes of violence by and against vegetables, which ultimately, Guanio-Uluru argues, lead the reader through a playful and multifaceted exploration of the moral standing of plants. In a different setting, Chapter 13, "The Vegetal Modality of Resistance in Children's Books by/for Indigenous Peoples in the Philippines" by Jose Monfred C. Sy, examines how the integral relationship of indigenous peoples of the Philippines to the land and to plants are reflected in a selection of indigenous children's books, forming part of an activist and preservationist stance.

The chapters of this collection thus wind their own paths through many of the issues raised in this introduction, exploring notions of kinship and agency, representation, and connection. The authors draw varying conclusions about the ultimate possibilities and limitations of engaging with plants through children's and YA literature, but affirm the vibrancy of the endeavour. Plants encountered in this collection include ferns, mulberry trees, gooseberry bushes, chamomiles, vegetables, guava trees, oak trees, roses, apple trees, rhubarb plants, willow trees, vines, forests, and magical plants, both vicious and benign. Poetry, picturebooks, YA and middle-grade novels, illustrated novels, and fantasy are explored. Future work might consider the plethora of richly illustrated non-fiction works currently being published for children. Children's and YA literature emerge as spaces of encounter with the vegetal world—spaces of engagement and transformation.

Notes

1 The term was originally proposed by Paul Crutzen but has its critics. Rejecting its anthropocentrism, Donna Haraway (2016) prefers the term the Capitalocene, because it more accurately diagnoses the causes of the present global predicament of global warming due to an over-taxation of natural resources. Edward O. Wilson terms the "Anthropocene world view" dangerous, because it is counterproductive to conservation efforts by suggesting that the damage has already been done and is irreversible (2016, p. 71–79).

2 A similar argument was furthered by Michael Pollan in *The Botany of Desire: A Plant's Eye View of the World* (2001).

References

Anonymous. (2013). *The dream of the rood*. In S. Greenblatt (Ed.) *The Norton anthology of English literature: The major authors*, vol. 1 (pp. 34–37). New York and London: Norton.

Barad, K. (2007). *Meeting the universe halfway: Quantum physics and the entanglement of matter and meaning*. Durham, NC: Duke University Press.

Bennett, J. (2010). *Vibrant matter: A political ecology of things*. Durham, NC: Duke University Press.

Bergthaller, H. (2006). "Trees are what everyone needs": The lorax, anthropocentrism, and the problem of mimesis. In C. Gersdorf & S. Mayer (Eds.), *Nature in literary and cultural studies: Transatlantic conversations on ecocriticism* (pp. 155–175). Amsterdam and New York: Rodopi.

Beskow. E. (1986). *Children of the forest* (2nd rev. ed.). Edinburgh: Floris Books. (Original work published 1910.)

Blyton, E. (2016). *The magic faraway tree collection: 3 Books in* 1. London: Egmont Books. (Original works publishes 1939, 1943, 1951.)

Bradford, C., Mallan, K., Stephens, J., & McCallum, R. (Eds.) (2008). *New world orders in contemporary children's literature: Utopian transformations.* Basingstoke and New York: Palgrave Macmillan.

Braidotti, R. (2013). *The posthuman.* Cambridge: Polity Press.

Buell, L. (1995). *The environmental imagination: Thoreau, nature writing, and the formation of American Culture.* Cambridge, MA: Harvard University Press.

Buell, L. (2005). *The future of environmental criticism.* Malden: Wiley-Blackwell.

Buell, L. (2014). Environmental writing for children: A selected reconnaissance of Heritages, emphases, horizons. In G. Garrard (Ed.), *The Oxford handbook of ecocriticism* (pp. 408–422). Oxford: Oxford University Press.

Burnett, F. H. (2012). *The secret garden.* London: Vintage. (Original work published 1911.)

Carson, R. (1962). *Silent* spring. New York: Houghton Mifflin Company.

Castro, I. E., & Clark, J. (2019). *Representing agency in popular culture: Children and youth on page, screen, and in-between.* Lanham, MD: Lexington.

Curry, A. (2013). *Environmental crisis in young adult fiction: A poetics of earth.* Basingstoke and New York: Palgrave Macmillan.

Cutter-Mackenzie, A., Payne, P., & Reid, A.(Eds.). (2011). *Experiencing environment and place through children's literature.* Abingdon: Routledge.

Darwin. C. (1875). Insectivorous plants. http://darwin-online.org.uk/content/frameset?itemID=F1217&viewtype=text&pageseq=1

Darwin, C. (1880). *The power of movement in plants.* London: J. Murray.

Darwin, E. (1796). *Zoonomia, Vol.* I *Or, the laws of organic life*, Section XIII, "Of vegetable animation". Retrieved from www.gutenberg.org/files/15707/15707-h/15707-h.htm

Deleuze, G., & Guattari, F. (1987). *A thousand plateaus: Capitalism and schizophrenia.* University of Minnesota Press. (Original work published 1980.)

Derrida, J. (2002). The animal that therefore I am (more to follow). *Critical Inquiry*, 28(2), 369–399.

Dobrin, S. I., & Kidd, K. B. (Eds.) (2004). *Wild things: Children's culture and ecocriticism.* Detroit: Wayne State University Press.

Frazer, J. (2009). *The golden bough: A study in magic and religion.* Oxford: Oxford University Press. (Original work published 1890.)

Gaard, G. (2008). Toward an ecopedagogy of children's environmental literature. *Green Theory & Praxis: The Journal of Ecopedagogy*, 4(2). https://doi.org/10.3903/gtp.2008.2.3

Gagliano, M. (2018). *Thus spoke the plant: A remarkable journey of groundbreaking scientific discoveries and personal encounters with plants.* Berkeley, CA: North Atlantic Books.

Gagliano, M., Ryan, J. C., & Vieira, P. (2017). *The language of plants: Science, philosophy, literature.* Minneapolis: University of Minnesota Press.

Glotfelty, C., & Fromm, H. (Eds.). (1996). *The ecocriticism reader: Landmarks in literary ecology*, (pp. xv–xxxvi). London: The University of Georgia Press.

Goga, N. (2017). I begynnelsen var treet. Økokritisk lesning av omformingen fra et stykke tre til gutt I Carlo Collodis *Le avventure di Pinocchio. Storia di un burattino* (1883). *Nordic Journal of ChildLit Aesthetics*, 8(1). doi:10.1080/20007493.2017.1308750

Goga, N. (2020). Økokritiske perspektiv på representasjoner av skog i skandinavisk barnelitteratur. In U. Nemeth & A. Nordenstam (Eds.), *Svensklärarföreningens ärsskrift 2019* (pp. 113–128). Stockholm: Svensklärarföreningen.

Goga, N., Guanio-Uluru, L., Hallås, B. O., & Nyrnes, A. (Eds.). (2018). *Ecocritical perspectives on children's texts and cultures: Nordic dialogues*. Basingstoke: Palgrave Macmillan.

Graulich, M. (1996). "Fascist fishes, power rangers and rape crisis feminists: Who gets to define?" *ISLE: Interdisciplinary Studies in Literature and the Environment*, 2(2), 93–98.

Griffiths A., & Denton T. (2011–) *The treehouse books*. London: Macmillan Books.

Guanio-Uluru, L. (2015). *Ethics and form in fantasy literature: Tolkien, Rowling and Meyer*. Basingstoke: Palgrave Macmillan.

Guanio-Uluru, L. (2018). Plant-human hybridity in the story world of Kubbe. In N. Goga, L. Guanio-Uluru, B. O. Hallås, & A. Nyrnes (Eds.), *Ecocritical perspectives on children's texts and cultures: Nordic dialogues* (pp. 125–140). Basingstoke: Palgrave Macmillan.

Guanio-Uluru, L. (2019). Imagining climate change: The representation of plants in three Nordic climate fictions for young adults. *Children's Literature in Education*. https://doi.org/10.1007/s10583-019-09387-4

Gubar, M. (2013). Risky business: Talking about children in children's literature criticism. Children's Literature Association Quarterly, 38(4), 450–457.

Gubar, M. (2016). The hermenuetics of recuperation: What a kinship-model approach to children's agency could do for children's literature and childhood studies. *Jeunesse: Young People, Texts, Cultures*, 8(1), 291–308.

Hall, M. (2011). *Plants as persons: A philosophical botany*. Albany: State University of New York Press.

Harari, Y. N. (2011). *A brief history of humankind*. London: Vintage.

Haraway, D. (1991). *Simians, cyborgs and women: The reinvention of nature*. New York: Routledge.

Haraway, D. (2008). *When species meet*. Minneapolis: University of Minnesota Press.

Haraway, D. (2016). *Staying with the trouble: Making kin in the chthulucene*. London: Duke University Press.

Hayles, N. K. (1999). *How we became posthuman: Virtual bodies in cybernetics, Literature and Informatics*. Chicago: University of Chicago Press.

Hayles, N. K. (2017). *Unthought: The power of the cognitive nonconscious*. Chicago: University of Chicago Press.

Henderson, B., Kennedy, M., & Chamberlin, C. (2004). Playing seriously with Dr. Seuss: A pedagogical response to *the lorax*. In S. I. Dobrin & K. B. Kidd (Eds.), *Wild things: Children's culture and ecocriticism* (pp. 128–148). Detroit: Wayne State University Press.

Heneghan, L. (2018). *Beasts at bedtime: Revealing the environmental wisdom in children's literature*. Chicago: University of Chicago Press.

Hines, M. (2004). "He made us very much like the flowers". Human/nature in Anglo-American literature. In S. I. Dobrin & K. B. Kidd (Eds.), *Wild things: Children's culture and ecocriticism* (pp. 16–30). Detroit: Wayne State University Press.

Iovino, S., & S. Oppermann (Eds.). (2014). *Material ecocriticism*. Bloomington: Indiana University Press.

Irigaray, L., & Marder, M. (2016). *Through vegetal being: Two philosophical perspectives*. New York: Columbia University Press.

Jaques, Z. (2013). Arboreal myths: Dryadic transformations, children's literature, and fantastic trees. In I. Gildenhard, & A. Zissos (Eds.), *Transformative change in western thought: A history of metamorphosis from Homer to Hollywood* (pp. 163–182). Oxford: Legenda.

Jaques, Z. (2015). *Posthumanism and children's literature: Animal, environment, Cyborg*. London: Routledge.

Kokkola, L. (2016). Becoming native? The wisdom of plants in Margaret Engle's the surrender tree. *International Research in children's literature*, 9 (1), 35–49.

Kokkola, L. (2017). Critical plant studies and children's Literature. In C. Beauvais & M. Nikolajeva (Eds.), *Edinburgh companion to children's literature* (pp. 274–280). Edinburgh: Edinburgh University Press.

Laist, R. (2013). *Plants and literature: Essays in critical plant studies*. Amsterdam: Rodopi.

Li, R., & Ryan, J. C. (2018). Sowing seeds: Phytocriticism and the botanical dimensions of Indonesian literature for children and young adults. In J. C. Ryan (Ed.), *Southeast Asian ecocriticism: Theories, practices, prospects* (pp. 229–251). Lanham, MD: Lexington.

Maitland, S. (2012). *Gossip from the forest: The tangled roots of our forests and fairytales*. London: Granta Publications.

Mancuso, S., & Viola, A. (2016). *Brilliant green: The surprising history and science of plant intelligence*. Washington: Island Press.

Marder, M. (2013). *Plant-thinking: A philosophy of vegetal life*. New York: Columbia University Press.

Marder, M. (2015). *The philosopher's plant: An intellectual herbarium*. New York: Columbia University Press.

Marder, M. (2016). *Grafts: Writings on plants*. Minneapolis, MN: Univocal.

Marshall, I. (1996). The lorax and the ecopolice. *ISLE: Interdisciplinary Studies in Literature and the Environment*, 2 (2) 85–92.

Massey, G., & Bradford, C. (2011). Children as ecocitizens: Ecocriticism and environmental texts. In K. Mallan & C. Bradford (Eds.), *Contemporary children's literature and film: Engaging with theory* (pp. 208–217). Basingstoke: Palgrave Macmillan.

Miller, A. (2012). The giving tree and environmental philosophy: Listening to deep ecology, feminism and trees. In P. R. Costello (Ed.), *Philosophy in children's literature* (pp. 251–266). Plymouth: Lexington Books.

Montgomery, L. M. (1994/1908). *Anne of green gables*. London: Penguin.

Næss, A. (1989). *Ecology, community and lifestyle: Outline of an ecosophy* (D. Rothenberg, Ed. and trans.). Cambridge: Cambridge University Press.

Nealon J. (2016). *Plant theory: Biopower and vegetable life*. Stanford: Stanford University Press.

Osborne, M. P. (1992–). The magic tree house series. New York: Random House.

Plumwood, V. (2002). *Environmental culture: The ecological crisis of reason*. London and New York: Routledge.

Pollan, M. (2001). *The botany of desire: A plant's eye view of the world.* New York: Random House.

Pollan, M. (2013). The intelligent plant. *The New Yorker*, 3 December. https://www.newyorker.com/magazine/2013/12/23/the-intelligent-plant

Ross, S. (1996). Response to 'The lorax and the ecopolice' by Ian Marshall. *ISLE: Interdisciplinary Studies in Literature and the Environment*, 2 (2) 99–104.

Rowling, J. K. (1997–2007). Harry potter series. London: Bloomsbury.

Ryan, J. C. (2016). "Tolkien's sonic trees and perfumed herbs: Plant intelligence in Middle-earth". In P. Vieira, M. Gagliano, and J. Ryan (Eds.), *The green thread: Dialogues with the vegetal world* (pp. 37–58). London and New York: Lexington Books.

Ryan, J. C. (2018). *Plants in contemporary poetry: Ecocriticism and the botanical imagination.* London and New York: Routledge.

Schramski, J. R., Gattie, D. K., & Brown, J. H. (2015). Human domination of the biosphere: Rapid discharge of the earth-space battery foretells the future of humankind. *Proceedings of the National Academy of Sciences*, 112(31) 9511–9517. doi:10.1073/pnas.1508353112

Sendak, M. (2000). *Where the wild things are.* London: Random House. (Original work published 1963).

Seuss, T. G. (2012). *The lorax.* New York: Harper Collins. (Original work published 1971).

Silverstein, S. (2014). *The giving tree.* New York: Harper Collins. (Original work published 1964.).

Singer, P. (2002). *Animal rights: A new ethics for our treatment of animals.* New York: Harper Collins. (Original work published 1975.)

Stone, C. D. (2010). *Should trees have moral standing?: Law, morality and the environment.* Oxford: Oxford University Press. (Original work published 1972)

Sturluson, S. (2005). *The prose edda* (Jessie L. Byock, Ed.) London: Penguin Classics.

Taylor, M. (2016). *The song of the trees.* New York: Penguin. (Original work published 1975.)

Tolkien. J. R. R. (2005). *The lord of the Rings*: 50 Anniversary Edition. London: HarperCollins Publishers. (Original work published 1954–1955.)

Wandersee, J. H., & Schussler, E. E. (1999). Preventing plant blindness. *The American Biology Teacher*, 61(2), 84–86.

Wandersee, J. H., & Clary, R. E. (2006). Advances in research toward a theory of plant blindness. In *Proceedings of the 6th International Congress on Education in Botanic Gardens at Oxford University.* London: Botanic Gardens Conservation International.

Wilson, E. O. (2016). *Half-earth: Our planet's fight for life.* New York: Liveright Publishing Corporation.

Wolfe, C. (2010). *What is posthumanism?*Minneapolis, MA: University of Minnesota Press.

Wyndham, J. (1951/2003). *The day of the Triffids.* Modern Library.

Part I

Botanical Fascinations

1 A Relational Poetics of Plant–Human Interactions

Contrasting the Picturebooks of Cicely Mary Barker and Elsa Beskow[1]

Terri Doughty

The United Nations declared 2020 the Year of Plant Health (http://ww.fao.org/plant-health-2020/hom/en/), recognizing the importance globally of the need to protect plant species diversity and habitat. It can be challenging, however, to mobilize humans to promote plant health. This is perhaps connected to "plant blindness", which Margaretha Häggström (2017) defines in part as "incapacity to see or observe plants in our environment" and "failure to understand plants' function on earth" (p. 91). I would elaborate that plant blindness also encompasses an inability to see the interconnections between humans and plants. Engendering more respectful, perhaps even cooperative, interactions between humans and plants requires a cognitive shift, for humans to reject species hierarchies and recognize that plants and humans are interconnected in webs of kinship, what Donna Haraway (2015) refers to as "rich multispecies assemblages" (p. 160). This chapter compares the flower fairies of Cicely Mary Barker with Elsa Beskow's plant–human hybrids to explore how they might be read as contributing to such a cognitive shift by re-visioning how humans live with plants.

As Lydia Kokkola (2017) notes, "children's literature has an important role to play in developing ecopedagogy" (p. 278). Likewise, Alexandra Lakind and Chessa Adsit-Morris (2018) state, "[w]e need alternative imaginaries" (p. 38) to transform ecological thinking. Both Barker's and Beskow's books can be read to generate an affective cognitive shift in children by providing such "alternative imaginaries", with the fantastic figures of flower fairies and hybrid plant people providing opportunities for empathetic engagement with the more-than-human. Recent research in environmental education has found that children can develop moral and affective understandings of plants (Villanueva, Villaroel, & Antón, 2018; Collado & Sorrel, 2019). Moreover, such understanding can derive from vicarious as well as physical experience of nature (Soga, Gaston, Yamaura, Kurisu, & Hanaki, 2016), which suggests that picturebooks can have a role in shaping children's awareness of and sense of relationship to plant others.

A relational poetics of plant–human interactions emerges from contrasting the works of Barker and Beskow. Working in the first half of the 20th century, both women created iconic images of flower fairies or plant people,

DOI: 10.4324/9781032066356-3

blurring the plant–human boundary. However, Barker's illustrations, although notable for their botanical accuracy (Laing, 1995, p. 38; Scott, 2007, p. 291), mostly represent the plant fairies as individuals and sometimes pairs, not as part of a broader community of plants and people, though this is occasionally addressed in accompanying verse. Beskow provides similar types of illustrations for Jeanna Oterdahl's *Blommornas bok* [*The Flowers' Book*], a book of poems about flowers, early in her career, but in some of her own books, such as *Blomsterfesten i Täppan* [*The Flowers' Festival*] and *Lasse Liten i Trädgården* [*Christopher's Garden*],[2] she not only represents plants living within complex communities, but she shows human children interacting in non-exploitive ways with the flora. Beskow has been acknowledged for creating a Swedish cultural identity rooted in nature (see Halldén, 1989; Westin, 2008; Borg and Ullström, 2017). I argue that she also models a relational poetics of plant–human connections.

Defining a Relational Poetics of Human–Plant Interactions

Environmental philosophers and educators have developed various concepts to describe how to think about human and more-than-human differences and connections. Affrica Taylor (2013) writes about the need to "resituate[d] childhoods within down-to-earth natureculture common worlds" (p. 116). She specifies "thinking through our everyday multispecies interactions and relations … [and] recognis[ing] our interdependencies and agencies" (2017, p. 67), in short decentring the human for interrelations that promote the well-being of all. Likewise, discussing children and urban nature, Karen Malone (2018) uses the trope of entanglement to describe the relationship between child and nature as akin to "the connections with/in the human body itself, between its 'own' cells and the trillions of microorganisms … that inhabit and compose it" (p. 87). As humans are already in/of nature, relations between humans and other species are by necessity complex and messy. The image of entanglement is an effective means for thinking about child–plant relationships. Since humans could not survive without plants, we are all "entangled in an assemblage of collective vulnerability" (Lakind & Adsit-Morris, 2018, p. 36). Haraway (2016) also uses tropes of entanglement: string figures with interconnected threads and knots, spider webs and tentacles that stretch across boundaries. It is her notion of making kin, though, that is most evocative for a relational poetics of plant–human interactions: "all earthlings are kin in the deepest sense, and it is past time to practice care of kinds-as-assemblages (not species one at a time)" (p. 103).

To enact such relationships, Haraway (2016) notes, we must embrace both responsibility and "response-ability" (114). That is, humans need not only to be accountable for making kin, but also to be accountable for being responsive to the more-than-human. This is the other key aspect of a relational poetics for plant–human interactions. Research has indicated the capacity of children to develop an affective relationship with plants.

Villanueva et al. (2018) found that most children by age seven were able to identify plants as living beings to be treated empathetically. Where such empathy is not found, Collado and Sorrel (2019) suggest this is caused by children not seeing plants as having affectivity (p. 46). This is where literature for children has the capacity to enhance child–plant relationships: seeing affective engagement with plants as living entities in imagined realities may foster children's emotional sense of kinship with and respect for them as sentient lifeforms with agency. Häggström (2017) states that to "deeply understand non-humans we have to share some of their lived experience and be part of their lived time so as to expand and synthesize our horizons" (p. 89). This sharing, challenging in daily living due to different life spans of species, can take place metaphorically within the pages of children's books. Some might argue that the anthropomorphic characteristics of Beskow's plant people, which appear as plant–human hybrids (whereas Barker's flower fairies are human–fairy hybrids), are anthropocentric. However, anthropomorphism can also be thought of as a form of analogic thinking that does not necessarily carry pro-human bias; indeed, anthropomorphism may be a necessary element in humans' capacity to form attachments with nature (Lumber, Richardson, & Sheffield, 2017). A relational poetics of plant–child relationships in picturebooks is rooted in texts generating a sense of affective kinship with plants in child readers. While I argue that such affective kinship is most clearly present in Elsa Beskow's work, there are gestures toward this in Cicely Mary Barker's work as well, but the latter's flower fairies are less obviously part of intra- and interspecies assemblages.

Meeting Plants: The Flower Fairies of Cicely Mary Barker

The association of plants with fairies is an old one in the English literary tradition. Two of Titania's attendants in Shakespeare's *A Midsummer Night's Dream* are Peaseblossom and Mustardseed. It was the Victorians, however, who established a prominent link between fairies and children (Purkiss, 2003, p. 220). Furthermore, there emerged in the latter half of the 19th century a body of literature that linked children with plants and both with fairies (Hines, 2004; Forsberg 2015). The connection between the three was based on the idea that the fairy could be a locus of wonder that would draw children's attention to the marvels of nature. As Maude Hines (2004) notes, though, such texts were usually intended to improve children, and if there were any relationships imagined between children and plants, they were hierarchical, shaped by "paternalist sympathy" (p. 23). The popularity of nature fairies advanced in the late-19th and early-20th centuries. Barker's biographer, Jane Laing (1995), points to models of plant/fairy/child intersections in the work of Australian Ida Rentoul Outhwaite, whose books and postcards of nature fairies were available in England, and English artist Margaret Tarrant, later a friend of Barker's, who painted "Do You Believe in Fairies" (1922) the year before Barker's first flower fairy book, *Flower*

Fairies of the Spring, was published (p. 31). The latter was followed by *Flower Fairies of the Summer* (1925) and *Flower Fairies of the Autumn* (1926). Although she produced other flower fairy books into the 1940s, I will take most of my examples from these first three.

Much has been made of Barker's botanically accurate depictions of plants in these books. Laing (1995) reports that Barker supplemented direct observation with secondary resources, consulting G. Clarke Nuttall's *Wild Flowers as They Grow* and experts from the Kew Botanical Gardens, even getting samples from the latter (p. 38). Barker was particularly influenced by the Pre-Raphaelites, modelling her artistic practice on their commitment to painting from life to capture "nature in minute detail" (Laing, 1995, pp. 35–36). She also painted her children/fairies from life, using children from her sister Dorothy's kindergarten school or children of neighbours as models. Each plant with its fairy is accompanied by a song presented on the facing page. Although the position of the image varies from recto to verso, the eye is always drawn first to the vivid illustration. Laing approvingly describes Barker's painting style as more "naturalistic" than that of her friend Tarrant (p. 39). More recently, Carole Scott (2007) has also praised Barker's work for eschewing the stylized floral designs popularized by the Arts and Crafts movement, calling it "passionately ecological, with an emphasis on botanical correctness and a strong message to children that the natural world be honored and cared for" (p. 291). Certainly, the realistically drawn plants with their accompanying fairies and the helpfully titled verses assist readers to identify different plant species and teach them useful information about the plants' seasonal lives. However, depicting plants in realistic detail is not sufficient to establish a relational poetics, for while Barker gestures in some of her verses to the plants' relationships with children or other flora and fauna, she does not create strong images of interactive, multispecies communities.

The images in Barker's first three fairy books generally feature single fairies/plants; even in the 1930s when she began featuring fairy duos, they still usually belong to the same plant, though in *Flower Fairies of the Garden*, the Polyanthus and Grape Hyacinth Fairies are depicted in a conversation.[3] Visually, there is not a strong sense of individual plants as part of a larger natural community; they seem to be presented as specimens for identification, with each fairy perched on or carrying a sample of its plant. The accompanying verses do, though, reference interspecies interactions. Some fairies/flowers speak of feeding other species, such as the White Clover Fairy (Summer), which invites the bees to take its "honey", and Groundsel Fairy (Spring), which describes the plant as a treat for the "dicky-birds". Moreover, some spring flower fairies speak of being heralded by birds, such as the blackbirds and starlings "shouting" about the daffodil. There are also some references to relationships between plants. The Bluebell Fairy (Spring) compares himself to the Primrose, Woodland King to her Queen. In addition, the Lords-and-Ladies Fairy (Spring) mentions his "kinsman, Arum-lily". The verses, then, gesture at locating plants within broader communities.

Similarly, while there are not any non-fairy children depicted in the images, there are implied interactions between plants and humans in the verses. The Willow Catkin Fairy (Spring) sings about "unkind" people who "trample" and "spoil the little trees, and shout". Likewise, the Horse Chestnut Fairy (Autumn) complains about becoming "black and blue" from getting hit with the stones little boys throw into the tree to shake down conkers. On a more positive note, the Daisy (Spring) and Buttercup (Summer) invite children to play with them, both directly addressing the implied child reader. The verses vary in address: some use the first person and address the child reader, while others implicitly address adults, as the Dandelion Fairy (Spring) does when taunting "Sillies" who are unable to eradicate the plant, despite "spades and hoes of iron", with iron notoriously dangerous to fairies as well as to plants. Not all first-person verses address the reading audience; some are descriptions of the plant, as in the song of the Harebell Fairy (Summer), which addresses the plant itself. Other verses, as for the Foxglove Fairy (Summer), present dialogue between the fairy and the plant. A very few, such as the Lords-and-Ladies Fairy (Autumn; some plants appear in multiple seasons), focalize the plant addressing the fairy. Despite the occasional references to children playing with plants, much of the content of the verse seems to be educational, whether moral—the implied caution to children who might damage plants with their undisciplined play—or scientific—teaching children to identify plants and their lore, similar to popular children's nature plays of the same period (Palmer, 2017). For instance, accompanying the song of the Old-Man's-Beard Fairy (Autumn) is a note: "Old-Man's-Beard is Wild Clematis; its flowers are called Traveller's Joy. This silky fluff belongs to its seeds". There are also warnings about plants that contain toxins, such as the Nightshade (Summer) and Lords-and-Ladies (Autumn). In short, while children are meant to engage with plant life, they do so from a hierarchical position that identifies plants as objects: they may be played with, they may be eaten, and they may have other uses for humans, as does the Elderberry (Autumn), the berries of which are used by "[c]ountry people" to make wine. Moreover, plant knowledge is beneficial to understand their uses and dangers. Aside from the implication in a few verses that plants can be harmed by humans, and vice versa, there is no affective kinship.

Likewise, the illustrations of the plants and their fairies do not effectively promote a notion of plants and humans in a multispecies assemblage. As mentioned above, the fairies were painted from child models, and they look like children, albeit with wings, ranging in age from toddlerhood to prepubescence. While most fairies/children wear recognizable items of human clothing, such as the short-sleeved tunic worn by the Wood-Sorrell Fairy (Spring), all have some element of the plant comprising their dress. For example, the Beechnut Fairy (Autumn) is clad in a tunic made from what appears to be a beech leaf and is wearing the spiky husk of a nut as a cap. Scott (2007) argues that "the fairy-children are merged with the flowers and trees they wear, which creates a sense of identification and love for the growing things around them—a bond of empathy" (p. 292). While this might be read as demonstrating the child and nature as one, I am not sure that using another species for clothing establishes a bond of empathy, as people do not necessarily empathize with the animals from which their shoes are made. Moreover, the presentation of the fairies/children sometimes enhances objectification, just as the plants themselves are sometimes objectified in the verses. These fairy embodiments of childhood's supposed natural innocence gaze at their flowers or at the viewer, staged for the reader's viewing pleasure: the Stitchwort Fairy (Spring) smiles with cocked head as she lifts the hem of her gown; the accompanying text assures readers that she has a prettier name, Starwort. The paintings are romantic, idealizing the fairy children at play within nature, rather than being of nature. They and "their" plant specimens may be together, but they are not part of a "rich multispecies assemblage" (Haraway, 2015, p. 160). Each flower fairy verse and painting pairing is discrete from the others, which frames and limits interactions. Indeed, far from presenting a relational poetics of plant/human interactions, the flower fairies today seem to appeal to nostalgic adults who wish to nurture rosy visions of innocent childhood (cf. Scott, 2007, p. 293).

Elsa Beskow's Relational Poetics of Plant/Human Interactions

Elsa Beskow began her career as an illustrator earlier than Barker began hers and also had more formal training than Barker. She studied art at a technical school and was particularly influenced by the Arts and Crafts movement; her professor encouraged students to incorporate Swedish nature in their ornamental drawings, depicting "pinecones, sticks, flowers, everything in the Art Nouveau style" (Schaffer, 2002, p. 148; translated by Alicia Sjölin). There is a clear connection between her work and Walter Crane's *Flora's Feast: A Fairy's Festival of Flowers* (1889). Another influence on Beskow is the Swedish illustrator Otillia Adelborg, especially her *Prinsarnes Blomsteralfabet* [*The Princes' Flower Alphabet*; 1892] (see Hammar, 2002, p. 445; Schaffer, 1995). Whereas Scott dismisses Beskow's plant illustrations as "picturesque" but not "ecological" (p. 291), there is a long tradition in Swedish criticism, dating from the 1950s, that identifies Beskow's work as botanically accurate and deeply significant to a national appreciation for nature (Westin, 2008, pp. 164–165). Nature features strongly in even Beskow's first independent work, published under her maiden name, Maartman, *Sagan om den Lilla, Lilla Gumman* [*The Tale of the Little, Little Old Woman*; 1897]: the illustrations are framed with floral designs, and on the page framed in catkins, a potted geranium is animated with a frowning face. I am most interested, though, in her books that represent plants and children in inter- and intraspecies relationships: here I focus on *Blomsterfesten i Täppan* [*The Flowers' Festival*; 1914], and *Lasse Liten i Trädgården* [*Christopher's Garden*; 1920]. There are indeed elements of the whimsical and the decorative in her illustrations, but her hybrid plant people suggest a strong kinship between plants and humans; moreover, her plants are still identifiable as distinct species. A naturalistic illustration style is not required for a relational poetics, rather there must be an affective relationship between humans and plants, and there must be a representation of plants as living beings with affectivity and agency. These books provide both.

A strong advantage of Beskow's books over Barker's in creating affective plant–human intra- and interspecies interactions comes from their form: whereas Barker presents a series of illustrations and verses, Beskow's books present stories, with space to build relations. As well, unlike Barker, who features children as fairies/fairies as children, with human children only referenced or addressed in the verses, Beskow integrates human children into her narratives. *The Flower's Festival* presents Lisa, a girl left alone on midsummer's eve, a liminal time when the veil between the human and other worlds is thinner; she meets the Midsummer Fairy, who explains that "every Midsummer [the flowers] can go where they like, though people can't see them". The fairy puts a drop of poppy juice into her eyes to allow Lisa to witness the flowers' midsummer party. Lisa does not directly interact with the plants, but she does, with her newly opened eyes, see that "the meadow [is] full of life". Beskow has said the story came to her when she was in a blooming meadow and felt the flowers "called to her" (Schaffer, 2002, p. 156; translated by Alicia Sjölin). Lisa observes the plants living in organized families and communities, not only sentient but possessing agency. This vision ends when Lisa falls asleep, but it is the non-human that has the last word, with a lullaby by the frog on the final page of the book, continuing its vibrant life without her. This is a story about a human child learning truly to see the world of plants.

In *Christopher's Garden*, the intermediary between the child and the plants is the avatar of September, who guides Christopher to meet the plants and trees living in his garden: the gooseberry boys and girls, a red currant girl, Old Man Blackcurrant, Mrs. Bramley (an apple tree), the Misses Plum, Mr. Pear, and Mrs. Cabbage, to name a few. Christopher interacts with all the plants, from playing with the plant children to debating which is the most important plant in the garden with Mrs. Cabbage. It is worth noting that in both of Beskow's books the plants are not fairies: they have no wings, and while they have anthropomorphized bodies, they are readily identifiable as plant people, whether through the use of clothing, as in Barker's books, or with stronger suggestions of hybridity, as with the Barley Boys and Mrs. Clover in *Flowers' Festival*, whose barley spikes and clover petals grow straight from their scalps. This anthropomorphism is not anthropocentric. Lisa is decentred in her witnessing of the plants' preparations for and enjoyment of the Flowers' Festival; she appears in only two black-and-white and one colour illustration in the first three pages of the book. Christopher is the focalizing character in *Christopher's Garden*, but he meets the plants on equal terms. In addition, neither Lisa nor Christopher has a transactional encounter with the plants: Lisa appreciates the plants' vibrant lives, and although Christopher is visiting a domestic garden, he is there to play and engage, not harvest. It is the plants that talk about human consumption of plants: the Gooseberry Boys will not give Christopher his ball because he has too greedily eaten berries in the past, and the red currant girl tells him the currants need to be picked. Mr. Pear does not offer fruit but tells Christopher his branches are too heavy and need to be given support by Christopher's father. There is a more complex relationship here than one in which humans simply take and plants give; there is a reciprocal element of care. In addition, Christopher appreciates Mrs. Bramley for her beauty, not just her edible apples. Similarly, he values the Dahlia, Aster, Marigolds and Nasturtiums. Indeed, at the end of the book, Christopher weeps at the thought of losing his new connection to plants and is consoled by a whisper from Mrs. Bramley: "We're all still here, even though you can't always see us". If Lisa has learned to see plants, Christopher has learned not only to see them but that the relationships among plants extend to people as well. Plants are so physiologically different from humans, it is difficult for adults, let alone children, to imagine kinship with them (Mancuso & Viola, 2015). By presenting anthropomorphized plants knowable to and even interacting with children, these books present plants and children as kin.

In addition to modelling these interspecies connections, Beskow's books present plants living within richly intertangled communities. There is a growing body of research arguing that plants are intelligent beings: they have senses, they communicate with other plants and even animals, and they are active and adaptable (Mancuso & Viola, 2015). Looking at the plant world in *The Flower's Festival*, there is a complex society, with the parade of plants organized according to biome. They come in families and are all ages, and within specific plant types there are more kinship groups. In addition, while the plants are at the centre of every illustration, there are also many images of various insects and birds, with part of the book comprising embedded plant stories told by various birds, demonstrating interspecies connections and elaborating on the plants' familial and social structures. The first three pages that present static scenes of Lisa gazing at garden and meadow plants are followed by a dynamic page with two image panels above and below the letterpress. While the top one shows Buttercup children having a snack and their mother cleaning the face of one child, the bottom panel shows Mrs. Beetroot and Mrs. Potato preparing food for the feast, as potato toddlers get underfoot and a Carrot woman, Radishes, and Pea-blossoms rush about with platters of food. The plants have a range of capacities that we think of as human, yet which are also properties of plants. For example, Mrs. Broadbean nurtures her babies in their pod, noting that they will grow up to go out in the world and be planted to have their own babies; plants reproduce just as humans do, and the bean "babies" are not anthropomorphized in the image. Vibrant plant families and communities are understood through their correspondence with human families and communities.

Beskow does replicate human hierarchies and gender roles in her plant world here, with the Rose Queen and the excluded weeds, and the female plants tasked with preparing the feast (Schaffer, 2002). However, it should be noted that she also slyly undermines these, especially class snobbery: haughty Madame Pansy is brought down a peg when she trips over her luxurious gown and needs her "small relations", the Violets, to help her. Most notably, the rebellious weeds in *The Flower's Festival* are presented sympathetically, with a certain degree of pleasure in their carnivalesque disruptions. In the painting of them, they glow with ruddy health, and a little Stickyweed cocks a snook, while a Senna sticks out his tongue. They boldly stare at the reader, two with crossed arms and two with arms akimbo. In the accompanying sparrow's song, the weeds are described as "always full of fun" and proclaim they will never be eradicated; this is born out, as despite a small Burdock boy getting in trouble for flinging burrs at the flowers, the Queen rejects violence against the weeds. As well, like in the story of Madam Pansy, the snobbiness of courtier plants is gently mocked, as Lady Fuschia "faint[s] away" at the actions of the disruptive Burdock. Beskow's treatment of plant hierarchies can be read as a commentary on human class hierarchies, but one could as easily read this as a commentary on human–plant hierarches: just as "superior" plants must accept the rights of weeds, humans should acknowledge the rights of plants.

In *Christopher's Garden*, as Hammar (2002) notes, the plants inhabit a much more equal world (pp. 466–467). There are no kings or queens. Even the hierarchy rooted in age, which requires children to be respectful to their elders, is undermined when Christopher argues with Mrs. Cabbage. Christopher rejects her insistence that status is rooted in utility. When Mrs. Cabbage rejects Mrs. Sunflower and Mrs. Dahlia as "frilly extras", Christopher places Mrs. Bramley above Mrs. Cabbage because she is a beautiful and useful plant. Mrs. Cabbage expresses her anger by throwing Christopher's ball into a flower bed, but the painting of her reinforces Christopher's views. Hammar (2002) sees Mrs. Cabbage as in part embodying a celebration of common, hearty meals (pp. 446–447); however, despite her formidable presence facing September and Christopher, just slightly off-centre and dominating the composition of the painting, just above and to the right of her head, there is a large caterpillar eating the corner of the painting. It encourages viewers to think that Mrs. Cabbage may not be as grand a lady as she imagines. The final colour image in the book reinforces the flattening of hierarchies. The painting centres September, singing the "Song of autumn", and facing the reader. Christopher shares the reader's perspective, as he, along with a Broadbean mother, Redcurrant child and Gooseberry child, has his back to reader, watching September. September is standing in front of an Aster that creates a halo effect around his head, and he is directly under the nurturing sun. Food plants and flowers mingle without distinction in a half circle around him. Mrs. Bramley and Mrs. Cabbage stand together, behind a Dahlia, a Nasturtium, a red currant girl and Old Man Black Currant. Not all plants must be useful to humans to deserve existence and even celebration.

Villanueva et al. (2018) observe that children younger than 7 or 8 years of age struggle to recognize plants as living things. This is where Beskow's picturebook fantasy worlds can play a role guiding children to understand plants as interconnected with each other and with humans. Barker's flower fairies series teaches children to see and know plants, but the intermediation of fairy children without any human children engaging with the plants, as well as the plants' isolation in discrete verses and images, does not create the same relationships as do Beskow's books. Haraway (2016) proclaims it is necessary for humans to "meet the unexpected, non-natal kin, and to strike up conversations" (p. 130), and this is what occurs in Beskow's *Christopher's Garden*. Moreover, here and in *The Flowers' Festival*, plants are represented as sentient, agential beings existing in complex inter- and intra-species relationships. A relational poetics of plant–human interactions may be facilitated by such works, encouraging child readers to develop affective understanding of and connections to plants.

Notes

1 Research for this chapter was supported by a Vancouver Island University Research Award. The author also acknowledges assistance from staff at the Swedish Institute for Children's Literature. Help with some translations from Swedish to English was provided by Alicia Sjölin.
2 Although I have read both Swedish and English editions of Beskow's books, I work here from the English translations as more accessible to a broader audience. From this point on, I refer to the books by their English titles. They are unpaginated.
3 Like most picturebooks, Barker's are unpaginated. Individual flower fairies are identified by the season/book in which they appear.

References

Adelborg, O. (1892). *Prinsarnes blomsteralfabet*. Stockholm: Albert Bonniers Förlag.

Barker, C. M. (1923). *Flower fairies of the spring*. London: Blackie.

Barker, C. M. (1925). *Flower fairies of the summer*. London: Blackie.

Barker, C. M. (1926). *Flower fairies of the autumn*. London: Blackie.

Barker, C. M. (1944). *Flower fairies of the garden*. London: Blackie.

Beskow, E. [as E. Maartman]. (1897). *Sagan om den lilla, lilla gumman*. Stockholm: Wahlström i Widstrand.

Beskow, E. [as E. Maartman]. [1897]. (1991). *The flowers' festival* (J. Tate, Trans.). Edinburgh: Floris Books. (Original work published in 1914)

Beskow, E. [as E. Maartman]. [1897]. (1992). *Christopher's garden* (J. Tate & P. Lawson, Trans.). Edinburgh: Floris Books. (Original work published in 1920)

Borg, C. B., & Ullström, M. (2017). The child in the forest: Performing the child in twentieth-century Swedish picture books. *LIR.journal*, 9, 7–25. Retrieved from http://ojs.ub.gu.se/ojs/index.php/LIRJ/article/view/4126/3377

Collado, S., & Sorrel, M. A. (2019). Children's environmental moral judgments: Variations according to type of victim and exposure to nature. *Journal of Environmental Psychology*, 62, 42–48. doi:10.1016/j.jenvp.2019.02.005.

Crane, W. (2002). *Flora's feast: A fairy's festival of flowers*. New York: Dover. (Original work published in 1889.)

Forsberg, L. (2015). Nature's invisibility: The Victorian microscope and the miniature fairy. *Victorian Studies*, 57(4), 638–666. Retrieved from https://www.jstor.org/stable/10.2978/victorianstudies.57.4.03

Häggström, M. (2017). An aesthetic and ethical perspective on art-based environmental education and sustainability from a phenomenological viewpoint. In O. Franck & C. Osbeck (Eds.), *Ethical literacies and education for sustainable development: Young people, subjectivity, and democratic participation* (pp. 85–103). New York: Palgrave Macmillan.

Halldén, G. (1989). *Barndomens skogar: Om barn i natur och barns natur*. Stockholm: Carlsson Bokförlag.

Hammar, S. (2002). *Solägget: Fantasi och Verklighet i Elsa Beskows Konst*. Stockholm: Albert Bonniers Forlag.

Haraway, D. (2015). Anthropocene, capitalocene, plantationocene, Chthulucene: Making kin. *Environmental Humanities*, 6, 159–165. Retrieved from environmentalhumanities.org/arch/vol6/6.7.pdf

Haraway, D. (2016). *Staying with the trouble: Making kin in the Chthulucene*. Durham, NC: Duke University Press.

Hines, M. (2004). "He made *us* very much like the flowers": Human/nature in nineteenth-century Anglo-American children's literature. In S. J. Dobrin & K. B. Kidd (Eds.), *Wild things: Children's culture and ecocriticism* (pp. 16–30). Detroit: Wayne State University Press.

Kokkola, L. (2017). Critical plant studies and children's literature. In C. Beauvais & M. Nikolajeva (Eds.), *The Edinburgh companion to children's literature* (pp. 274–280). Edinburgh: Edinburgh University Press.

Laing, J. (1995). *Cicely Mary Barker and her art*. London: Frederick Warne.

Lakind, A., & Adsit-Morris, C. (2018). Future child: Pedagogy and the post-anthropocene. *Journal of Childhood Studies*, 43(1), 30–43. doi:10.18357/jcs.v43i1.18263

Lumber, R., Richardson, M., & Sheffield, D. (2017). Beyond knowing nature: Contact, emotion, compassion, meaning, and beauty are pathways to nature connection. *PLoS ONE*, 12(5), e0177186. doi:10.1371/journal.pone.0177186

Malone, K. (2018). *Children in the Anthropocene: Rethinking sustainabilities and child friendliness in cities*. New York: Palgrave Macmillan.

Mancuso, S., & Viola, A. (2015). *Brilliant green: The surprising history and science of plant intelligence* (J. Benham, Trans.). Washington: Island Press. (Original work published in 2013)

Oterdahl, Jeanna. [1905]. *Blommornas bok*. Stockholm: Wilhelm Billes Bokförlag.

Palmer, A. (2017). "A pleasant way of teaching the little ones to recognise flowers": Instructional nature plays in early 20th century Britain. *Children's Literature in Education*, 48(4), 341–354. doi:10.1007/s10583-016-9283-5

Purkiss, D. (2003). *At the bottom of the garden: A dark history of fairies, hobgoblins, and other troublesome things*. New York: New York University Press.

Schaffer, B. (1995). *Barn och blomma*. In A. Banér (Ed.), *Ur barndomens historia* (pp. 197–211). Stockholm: Centrum för barnkulturforskning vid Stockholms universitet.

Schaffer, B. (2002). "Från skog till hav—naturgestaltningen hos Elsa Beskow." In U. Cederlöf (Ed.), *Elsa Beskow, vår barndoms bildskatt* (pp. 147–167). Stockholm: Nationalmuseum.

Scott, C. (2007). Ecological dress: Art, pedagogy, and ambiguity in the work of C. M. Barker. In C. Kuhn & C. Carlson (Eds.), *Styling texts: Dress and fashion in literature* (pp. 291–308). Amherst, NY: Cambria Press.

Soga, M., Gaston, K. J., Yamaura, Y., Kurisu, K., & Hanaki, K. (2016). Both direct and vicarious experiences of nature affect children's willingness to conserve biodiversity. *International Journal of Environmental Research and Public Health*, 13, 529. doi:10.3390/ ijerph13060529

Taylor, A. (2013). *Reconfiguring the natures of childhood*. New York: Routledge.

Taylor, A. (2017). Cosmopolitics of place: Towards urban multispecies living in precarious times. In K. Malone, S Truong, & T. Gray (Eds.), *Reimagining sustainability in precarious times* (pp. 61–75). Singapore: Springer.

Villanueva, X., Villaroel, J. D., & Antón, A. (2018). Environmental awareness and its relationship with the concept of the living being: A longitudinal study. *Sustainability*, 10, 2358. doi:10.3390/su10072358

Westin, B. (2008). "Beskows biotoper: Skogen och djuren." In J. Hansson & G. Eng (Eds.), *Vänbok till Sonja Svensson* (pp. 163–169). Stockholm: Opal.

2 Stamens and Pistils in the Same Flower

Queer Posthuman Performativity of Plants in the Finnish Fairy Tale "Pessi ja Illusia"

Katri Aholainen

Pessi ja Illusia is a literary fairy tale book by Finnish author Yrjö Kokko (1903–1977), which was published in Finland in 1944. Kokko wrote the book while on the front during the Continuation War (1941–1944), when Finland was at war with the Soviet Union. *Pessi ja Illusia* tells the story of Illusia, a fairy who has run away from her home in the rainbow to the forest, where she meets a troll named Pessi. Pessi introduces her, and the reader, to creatures of the forest, from plants and insects to birds and small mammals. The story is multi–layered, since it consists of the frame story narrated by a human narrator, the adventurous main storyline of Illusia the fairy and Pessi the troll, and embedded narratives on different animals and plants. These different stories, alongside photographs of plants and animals taken by the author, paint a picture of the Finnish forest as being full of many different ways of living. *Pessi ja Illusia* became popular among readers, and it has retained its place in the hearts of Finnish audience to this day: the story has been adapted into a ballet, couple of movies, and countless plays in professional and amateur theatres. It has also been translated into several languages.

At its time of publication in 1944, *Pessi ja Illusia* offered its readers an escape from the horrors of war into the fantastical world of a fairy tale. Its representations of nature have been interpreted as allegories of Finnish society, since there is apparent anthropomorphism in non-human animal and plant characters who, for example, go to university, get married, and discuss politics and religion (Lassila, 1999, p. 27; Saukkoriipi, 2003, Simola, 2013). In Western literature, plants are often represented as parts of the milieu in which the actual story with human or animal characters happens (Gagliano, Ryan & Vieira, 2017, p. x). However, in the worlds of fairy tales, agency is not limited to humans. It is not unusual to find intentional and active non-human characters in fairy tales, be they animals, plants, or even inanimate objects. For example, Zacharias Topelius (1818–1898) and Anni Swan (1875–1958), whose works have been hugely influential in Finnish literature for children, wrote tales that include plants as main characters or otherwise active participants of the story.[1]

DOI: 10.4324/9781032066356-4

The trope of speaking animals and plants has been criticized for anthropocentrism, but it has also been read as an emancipating move to give a voice to nature (Melkas, 2011a, p. 128). John C. Ryan develops this discussion by introducing the concept of the ecological plant voice. By shifting the focus from a voice as a purely auditory phenomenon, he aims to understand the plant voice as a plant's material relationship with its environment. (Ryan, 2017, pp. 275–276, 283.) In this chapter, following Ryan's suggestions, I understand speaking plant characters in *Pessi ja Illusia* as being "manifested through vegetal presence and human recognition of it" (Ryan, 2017, p. 291) I join Monica Gagliano, John C. Ryan and Patrícia Vieira as they propose that the textual plants always come into being in "dynamic relation to actual, living flora". (Gagliano, Ryan & Vieira, 2017, p. xvi). Hence, even the most metaphorical literary discourse on plants has to do with the material reality of the vegetal (Gagliano, Ryan & Vieira, 2017, p. xvi). The plant characters of *Pessi ja Illusia* derive not only from the imagination of the author or the socially constructed reality of people, but also from actual, material plants (see Ryan, 2017, p. 276).

Expressions of concern for nature have been presented in Finnish children's literature since the 19th century. The above-mentioned Topelius and Swan included ideas of animal and environmental protection in their tales. (Laakso, Lahtinen & Heikkilä–Halttunen, 2011, p. 9, 20; Kirstinä, 2011; Melkas, 2011b, p. 98). *Pessi ja Illusia* draws on this tradition but seeks to take the literary fairy tale in a more biologically accurate direction (Saukkoriipi, 2003, p. 162). Kokko had an education in veterinary science, and in his book *Alli. Jäämeren lintu* (1966), *Pessi ja Illusia* is referred to as "a biological fairy tale" (Kokko, 1966, 18). Taking this description as a starting point for my analysis, I ask in this chapter: how are biological knowledge and the traditions of fairy tales entangled in the plant characters of *Pessi ja Illusia*? Drawing on the feminist physicist Karen Barad's agential realism,[2] I read biology and fairy tales as material-discursive practices of boundary-making through which plant characters emerge.

These practices are performative, since they simultaneously produce and constrain the possibilities of agency. In theories of performativity[3] the focus of the analysis is shifted from actors to those acts that produce subjectification or "thingification".[4] This means that there are not assumed to be any fixed subjects or entities behind actions, but that these subjects or entities are produced through repetitive acting. As a Judith Butler (2002, p. 181) states, "there need not be a 'doer behind the deed', but that the 'doer' is variably constructed in and through the deed".

Karen Barad takes into consideration the ontological implications of these ideas when developing a posthuman elaboration of performativity. I find Barad's notion particularly useful for reading non-human characters, since it allows us to see matter as "an active participant in the world's becoming" (Barad, 2003, p. 803). In this sense, performative acts are not only forms of human discursivity, but ongoing material-discursive practices of boundary-making in which "part of the world makes itself intelligible to some other part" (Barad, 2007, p. 207, see also p. 185). Drawing on Barad's agential realism, and on human–plant studies as "a framework for conceptualizing plants as active partners in knowledge production and cultural practices" (Ryan, 2012, p. 115), I read the plant characters of *Pessi ja Illusia* as material-discursive phenomena.

In agential realist accounts, *a phenomenon* is a primary ontological unit which refers to the inseparability of intra-acting agencies (Barad, 2007, p. 141 and p. 333). Phenomena in agential realism do not refer to Kantian or the phenomenologist's phenomena (Barad, 2007, p. 140 note 18), but the term draws on quantum physicist Niels Bohr's use of phenomenon to describe "particular instances of wholeness" (Barad, 2007, p. 119); that is, the ontological inseparability of the measured object and the apparatus used to measure it. In agential realism, phenomena are material-discursive, which means that "no priority is given to either materiality or discursivity; neither one stands outside the other" (Barad, 2007, p. 177). As phenomena, understood in this sense, plant characters emerge through human and non-human agencies.[5] I seek to read materiality and discursivity through each other, not in isolation (Iovino & Oppermann, 2014, p. 5): how do plant characters emerge through performative entanglements of fairy tale traditions, biological knowledge and the material reality of plants?

Queering Sexual Reproduction in a Fairy Tale

The Finnish literary scholar Kukku Melkas (2011a, p. 130) has noted that there are numerous references to sexuality in *Pessi ja Illusia*, which are tightly linked to sexual reproduction. Many heteronormative non-human animal couples, especially birds, are introduced to the reader, with females sitting in nests while males are described as the providers of the family. The narrator does not spare his views on gender: for example, "But what can you do with girls anyway. They always behave so strangely." ("Mutta minkäpä noille tytöille yleensä voi. Nehän käyttäytyvät aina niin omituisesti.") (Kokko, 1944, pp. 21–22) and "Mrs. Redstart seemed to have a sense of humor, and that is a very rare gift for women." ("Rouva Leppälinnulla näytti olevan huumorintajua, ja se on hyvin harvinainen lahja naisilla.")[6] (Kokko, 1944, p. 92). As Melkas (2011a, p. 132) points out, in *Pessi ja Illusia*, women are narrated as caring mothers and men as fathers who bravely protect their families.

Melkas (2011a, p. 132) reads *Pessi ja Illusia* in the light of the historical context of war, when gender order has been said to be in "chaos" as men were at war and women on the home front had to take care of work that was traditionally seen as men's duty. In this context, Melkas writes, the biological views in *Pessi ja Illusia* serve to organize the chaos by naturalizing the binary gender order. On the other hand, Melkas (2011a, p. 131) notes that the biological perspective in *Pessi ja Illusia* seems to resist some aspects of the social order of 1940s Finland, since it seeks to represent sexuality as a natural instinct, not something to avoid. To some extent, I agree with Melkas' reading, but I still think that there is more to the biology in *Pessi ja Illusia*. In this chapter, I suggest that focusing on plant characters and reading them as material-discursive phenomena, as mentioned before, can open up ways to see biology differently, not only as a naturalizing force but as an opening to plants' queer performativity.

Lewis C. Seifert (2015, p. 18) points out in the introduction to a special issue of *Marvels and Tales* on queer fairy tale studies, that even though "fairy tales have been—and continue to be—used to enshrine normative heterosexual love", the link between the fairy tale and heteronormativity is not inevitable. To disrupt this linking, scholars need to queer fairy tales, which means, according to Seifert (2015, p. 16), to accentuate "what departs from normative social expectations about gender and sexuality, thus exposing the notions of 'normal' gender and sexual identities as myths". As performativity theories suggest, the focus of my analysis is on actions that produce one thing as normative and others as excluded from it.

To consider—and to queer—these ongoing performative negotiations of norms, "naturality", and the powers enacting them, I will focus in this chapter on descriptions of the sexual reproduction of plants. Emily Martin argued in 1991 (p. 485) that "the picture of egg and sperm drawn in popular as well as scientific accounts of reproductive biology relies on stereotypes central to our cultural definitions of male and female". Although in my performative analysis I see dynamics of scientific discourses and social order as more complex than the word "relying" would suggest, Martin's argument offers an interesting starting point for analysing descriptions of sexual reproduction in *Pessi ja Illusia*. How are the production of biological knowledge on sexual reproduction and the binary gender order material–discursively entwined in plant characters in *Pessi ja Illusia*? And, on the other hand, how can fairy tale plants queer these performative practices?

In this sense, reading the queerness of plant characters in a fairy tale book is not to interpret them as representations of some assumedly inherent, essential queerness, nor is it to glorify the agency of plants as enacting some kind of "natural" resistance against heteronormativity, but rather to make "the point that their very 'species being', as it were, makes explicit queering of 'identity' and relationality", as Barad (2011, p. 126) writes. Thus, to scrutinize plant characters' queer agencies is to take into consideration how they participate in practices of boundary-making and how this participation can make visible the performative repetition that produces inclusions and exclusions.

Plants' Ecological Entanglements with Insects

Some of the first creatures Illusia meets after arriving in the forest are Willow and Bumblebee. Pessi tells Illusia that Bumblebee works as a postman for the

Willows. He explains that Mr. Willow has bags of nectar to tempt Bumblebee, and while Bumblebee is collecting nectar, Mr. Willow attaches his letter to Bumblebee, who then delivers it to Ms. Willow. The Willows are said to write about love in their letters.

Pessi's explanation is, of course, an anthropomorphic description of insect-assisted pollination. It is notable that Kokko has chosen the willow as an example of the sexual reproduction practices of plants, since "Ms. Willows and Mr. Willows live in different shrubs" ("Pajuneidit ja herrat näet asuvat eri pensaissa") (Kokko, 1944, p. 33). In botanical terms, willows are dioecious, which means that some individual organisms of the plant have stamens in their flowers and are defined as male, while others have pistils and are defined as female. The difference between male and female willows is emphasized when Pessi tells Illusia that "Mr. Willows don't care about each other's letters" ("Pajuherrat eivät välitä toistensa kirjeistä") (Kokko, 1944, p. 32). Choosing a dioecious plant as an example to teach Illusia and readers about sexual reproduction makes it possible to describe the said process in heteronormative terms.

However, when considering the sexual reproduction of willows, it is apparent that the description does not really fit the idea of sexual practices as a private act between one male individual and one female individual. Willows need other organisms, other species, to accomplish fertilization. The sexual reproduction of willows is ecologically entangled with insects. The ecological approach, which aims to understand "the interactions of individuals with their surrounding physical environment and the wider community of organisms living within it" (Gagliano, Ryan & Vieira, 2017, xiii), helps to reframe literary descriptions of the sexual reproduction of dioecious willows differently than a heteronormative understanding would suggest.

In Barad's agential realist terms, the agencies of willows and bumblebees are ontologically entangled, as they emerge through *intra-action*. Barad proposes the concept of intra-action as an alternative to the more common term interaction. When working with the concept of interaction, it is presumed that there exist individual entities which can affect each other. In this case, the entities precede the relations in which they engage. Intra-action instead refers to the idea that "relata do not preexist relations" (Barad, 2007, p. 140). The world or reality is not composed of individual entities or "things–behind–phenomena" (Barad, 2007, p. 140) but of relata-within-phenomena. The willows and bumblebees in *Pessi ja Illusia* do not pre-exist the relation they engage in. Their agencies emerge through each other, ontologically entangled in intra-action. It is precisely this co-evolution of plants and other organisms on which Randy Laist is relying when stating that "flowers and fruits are semantic in their being" (2013, p. 14).

Analysis of willows and bumblebees shows that even when considering dioecious plants, the practices of sexual reproduction do not occur only between male and female individuals whose genetic material combines in fertilization. Flower and insect characters' intra-active becoming–with shows that material-discursive practices of sexual reproduction can be entanglements of multispecies agencies. This relationship between flowers and insects is represented in many terms in *Pessi ja Illusia*. It is a mutualist cooperation that benefits both parties, but it is referred to also as friendship when the narrator states that "Insects are plants' best friends" ("Hyönteiset ovat kasvien parhaita ystäviä") (Kokko, 1944, p. 64). In some cases, the relationship can also be romantic, like in princess fairy tales:

> This time the butterfly prince's falling in love was serious. He got married to Cornflower. Cornflower became a princess and she got the new and fancy noble name Centaurea.
>
> Tällä kertaa perhosprinssin rakastuminen oli vakavaa laatua. Se meni naimisiin ruiskaunokin kanssa. Ruiskaunokista tuli prinsessa ja se sai uuden hienon aatelisnimen Centaurea.
>
> (Kokko, 1944, p. 124)

Zacharias Topelius also wrote about the relationship of flower and butterfly as a love story in his tale "Hvitsippan" (1866),[7] although this tale of an Anemone flower and the Apollo butterfly differs from Kokko's story in its tragic tone. Nonetheless, anthropomorphic characters in Kokko's and Topelius' tales allow us to think about the relationship between plants and insects not only in terms of benefit or survival, but in terms of emotional commitments and affective entanglements.

As material-discursive phenomena who emerge through human and nonhuman agencies, plant characters open up spaces to see sexual reproduction as a set of practices that go beyond the heteronormative image of sexual reproduction as an interaction between one female and one male individual. An ecological approach helps to reframe the agencies of sexual reproduction, and thus to show that heteronormative ideas are productions of a performative repetition of norms, not the determined consequences of a fixed state of nature. It is important to notice that the power of performatives is twofold: at the same time as they produce one thing as "normal" and others as not, they hide their own action in order to present determinations made by them as inherent and thus unavoidable features of reality (See Butler, 2002, pp. 5, 173–174). As the willows and bumblebees in *Pessi ja Illusia* show, the plant and insect characters can disrupt and queer this continuous producing and hiding of the norms.

"Mr.–Mrs. Rowan"

Toward the end of the story, the tension grows, as Pessi heroically fights a weasel named Martes who tries to hunt him and Illusia. Martes dies and Pessi is injured. He falls asleep and could sleep for the whole winter, as trolls do, but this time he has not eaten enough to survive hibernation. Illusia needs to get him nourishment, and since it is a snowy winter, there are not so many options. Her friend, Mrs. Field Mouse, recommends rowan berries, and so Illusia goes to Rowan.

> –Good afternoon, Mrs. Rowan, said Illusia politely.
> –Good afternoon, answered Rowan hoarsely. – Mrs. or Mr., however you want, Rowan continued. – I bear stamen and pistil flowers in the same tree, or more precisely; the stamens and pistils are in the same flower.
> –Hyvä päivää, rouva Pihlaja, sanoi Illusia kohteliaasti.
> –Hyvää päivää, vastasi pihlaja käheästi. – Rouva tai herra, miten vain tahdotte, pihlaja jatkoi. – Minulla on kyllä hede ja emikukat samassa puussa, tai tarkemmin sanottuna; heteet ja emit ovat samassa kukassa.
> (Kokko, 1944, p. 194)

At this point of the story, Illusia has met and heard stories about many plants, dioecious and monoecious, who are referred to mainly as male or female. Rowan is the only character that comments on having both stamens and pistils; that is, some attributes defined as female and other attributes defined as male, in the same individual organism.

Rowan's reaction to Illusia calling them Mrs. is ambivalent. They[8] do not seem to care too much about what they get called: "Mrs. or Mr., however you want"; but at the same time, they do want to point out that they are not really "Mrs". and to tell Illusia (and the readers) about their monoecy. "However you want" passes the problem of finding the correct words to Illusia and shows that it is not really Rowan themself who needs these kinds of determinations, as they are perfectly happy with having "pistils and stamens in the same flower". Illusia then solves the problem by apologizing for her ignorance and addressing Rowan as "Mr.–Mrs. Rowan". Then Illusia and Rowan continue to talk about berries, and leave the issue of stamens and pistils aside.

Lewis C. Seifert sees the coexistence of realism and fantasy characteristic of fairy tales as a paradox that can open up space for queerness (Seifert, 2015, p. 19). The point of this simultaneity helps in reading Mr.–Mrs. Rowan as an interplay of anthropomorphic fairy tale traditions and Kokko's commitment to biological knowledge. In the simultaneous contexts (or intra-action) of a fairy tale and biology, the result is "Mr.–Mrs. Rowan", a non-human character who refuses to be determined by Illusia's normative performative act of addressing them as Mrs. The place in which Mr.–Mrs. Rowan emerges is particularly a site of a fairy tale in its possibility to mix realistic and fantastical elements.

Performativity does not mean free will to choose how to perform one's identity. Instead, it means that identities are constructed in the field of performative repetition that produces some performatives as normative and others as excluded from the "normal" (Butler, 1993, p. 15). The character of

Rowan makes explicit this interplay of repeating the norms conventionally and making difference within them. Rowan must comment on their relation to the binary gender order even as they refuse to be determined by it. As a material-discursive phenomenon, the character of Rowan cannot detach themself entirely from the discourse on genders, but this does not mean that those norms succeed in defining them completely. Mr.–Mrs. Rowan's queer agency does not come from outside of the norms[9] but emerges through repeating them differently.

Refusal of the binary gender order of human social norms is not the only way in which Rowan challenges the anthropocentric idea of nature as an object waiting to be interpreted and determined by humans.

> Also humans say that—Rowan started coughing hard—when the rowan bears a lot of berries, it predicts war, continued Rowan. – Do you notice how self–centered he is! The human says that it predicts. See, he doesn't dare to say that those berries which have grown in a tree this hard make the war. He is so self–centered that he thinks that he controls everything in the world, that he is not subject to the same laws of nature, cause and effect, as the rest of the world. He doesn't know that we rowans take revenge on him, not only on our own behalf, but also for the whole of nature, from which he has separated himself.
>
> Ihmisetkin sanovat, että—pihlaja alkoi rykiä ankarasti, —kun pihlajassa on paljon marjoja, se ennustaa sotaa, jatkoi pihlaja puhettaan. – Huomaatko miten itserakas hän on! Ihminen sanoo, että se ennustaa. Katso hän ei uskalla sanoa, että ne marjat, jotka ovat kasvaneet näin kovassa puussa, tekevät sodan. Hän on niin itserakas, että hän luulee hallitsevansa kaikkea maailmassa, ettei hän ole samojen luonnonlakien, syyn ja seurauksen lakien alainen, kuin muutkin maailmassa. Hän ei tiedä, että me pihlajat kostamme hänelle, ei yksistään omasta, vaan myöskin koko luonnon puolesta, josta hän on eronnut.
>
> (Kokko, 1944, p. 195)

Rowan accuses people of arrogantly thinking that the only agency that matters—and even gets to be seen as agency in the first place—is human. Anthropocentric ideas of agency are turned around as Rowan claims that their berries are not a tool for people to predict the future, but actually cause war. Rowan reminds readers that human agency, even when trying to cut itself off from the non-human world, is entwined with and emerges through non-human agencies.

Rowan says that humans are "subject to the same laws of nature, cause and effect", which refers to Newtonian classical mechanics and its understanding of laws of nature. However, reading in an agential realist account allows us to see this shift in agency suggested by Rowan as bringing forth the indeterminacy of causality. Determining some things as causes and others as effects is an ongoing process enacted in material-discursive practices of including and excluding within–phenomena, which means that it is not an inevitable interpretation of inherent properties of entities (See Barad, 2003, 815; 2007, pp. 175–177; 2011, p. 136). As Rowan states, plants do not just wait for humans to make them meaningful. As performative agents, they participate in the practices in which they come to matter (Barad, 2007, pp. 151–153). Erin James makes a similar point on Ursula Le Guin's short story "Direction of the Road", which is narrated by an oak. According to James, the oak narrator leads readers to reimagine "laws of physics from a completely unfamiliar perspective, in which a plant lies at the heart" (James, 2017, p. 261). So does Rowan in *Pessi ja Illusia* when proposing a viewpoint that differs from anthropocentric understandings of nature and physics.

This opening up to different determinations of causes and effects is part of the queer performativity of "Mr.–Mrs. Rowan". As a fictional character who emerges through the intra-action of the material reality of rowans and the social gender norms of 1940s Finland, Rowan makes visible the ongoing performative repetition that produces binary gender. By questioning the ideas of causality, they show readers that the binary understanding of genders is not an unavoidable effect of sexual reproduction, but is performatively produced in material-discursive practices, such as repeatedly addressing some bodies as Mr. and others as Mrs. The ecological entanglement of willows and bumblebees reframes agencies of sexual reproduction *beyond* individuals of the species, while Mr.–Mrs. Rowan directs attention to the indeterminacy *of* the individual character.

Conclusions

A commitment to biological knowledge made Kokko write—especially when considering plants—characters and descriptions of sexual practices that challenged the heteronormative binary gender order of 1940s Finland. However, still there are also undeniably heterosexist descriptions and narrations in *Pessi ja Illusia*. From the agential realist point of view, which takes account of phenomena instead of entities, it does not make sense to determine whether, after all, *Pessi ja Illusia* is a queer or a heteronormative fairy tale. Instead, I have proposed in this chapter an understanding of the relation between biology and fairy tale in *Pessi ja Illusia* as an entanglement of different material-discursive forces that repeatedly negotiate the boundaries of subject and object, cause and effect, human and non-human. My analysis shows that in *Pessi ja Illusia*, biology is not only a tool to naturalize particular social agencies or identities over others, but instead, when intersecting with fairy tale, biology can show the way to opening up the established social orders for reconsideration.

These reconfigurations occur by re-framing the agencies that participate in material-discursive processes. The ecologically oriented analysis of willows and bumblebees shows that sexual reproduction practices are not an issue of separate entities or subjects but emerge through the intra-actions of the material-discursive, multispecies world. The character of Mr.–Mrs. Rowan challenges binary gender norms by interrupting their performative repetition and bringing forth the indeterminacy of causalities. In *Pessi ja Illusia*, the entanglement of biological knowledge, anthropomorphic fairy tales, and social norms makes visible the fact that normative ideas of binary gender are not an unavoidable consequence of sexual reproduction, but are produced as such in performative practices of inclusion and exclusion. And as my reading of *Pessi ja Illusia* shows, a fairy tale, as a set of intra-actions of human and non-human agencies, can become a place in which to queer these material-discursive practices of boundary-making.

Notes

1 See, for example, Topelius: "Hvitsippan" (1866), "Björkens Stora planer i mössörontiden" (1867) and "Linet" (1884); Swan: "Orava ja näre" and "Älykäs haukka" (1933).
2 Barad, 2007, especially chapter 4, "Agential realism: How Material-Discursive Practices Matter", pp. 132–185.
3 The varieties of performativity theories can be traced back to J. L. Austin's speech act theory and its linguistic followings. Barad has developed posthumanist notion of performativity by reading Judith Butler's work on performativity and sex/gender, and Michel Foucault's work on power alongside the physics–philosophy of the quantum physicist Niels Bohr. (Barad, 2003, 2007.) In this chapter, I mainly rely on Barad's posthumanist notion, but I find it particularly useful to read Barad with the feminist Judith Butler's writings on performativity. For an introduction to performativity theories and feminism, see, for example, Lloyd, 2016.
4 "*Thingification*—the turning of relations into 'things', 'entities', 'relata'" (Barad, 2003, p. 812).
5 When I write about human and non-human agencies, I am not referring to them as inherently separate actions. In agential realist terms, "human" or "non-human" is something that emerges through intra-actions and is then agentially cut (Barad 2011, p. 125) into human and non-human.
6 The translations of *Pessi ja Illusia* in this chapter are mine.
7 I thank the anonymous peer-reviewer for introducing me to this tale of Topelius.
8 Kokko refers to non-human characters with the pronoun "se" = "it", since in Finnish, the third person singular "hän" (which is gender-neutral) is used only when referring to people. In this chapter, I have decided to refer to Rowan with the singular "they" to respect their non-binary nature.
9 "The paradox of subjectivication (*assujetissement*) is precisely that the subject who would resist such norms is itself enabled, if not produced, by such norms. [...] it does locate agency as a reiterative or rearticulatory practice, immanent to power, and not a relation of external opposition to power" (Butler 1993, p. 15).

References

Barad, K. (2003). Posthumanist performativity: Toward an understanding of how matter comes to matter. *Signs: Journal of Women in Culture and Society*, 28(3), 801–831. Retrieved 28 October 2020 from www.jstor.org/stable/10.1086/345321

Barad, K. (2007). *Meeting the universe halfway*. Durham and London: Duke University Press.

Barad, K. (2011). Nature's queer performativity. *Qui Parle: Critical Humanities and Social Sciences*, 19(2), 121–158. Retrieved 28 October 2020 from www.jstor.org/stable/10.5250/quiparle.19.2.0121

Butler, J. (1993). *Bodies that matter: On the discursive limits of "sex"*. New York and London: Routledge.

Butler, J. (2002). *Gender trouble* (10th anniversary ed.). London: Routledge. Retrieved 28 October 2020 from ProQuest Ebook Central https://ebookcentral.proquest.com

Gagliano, M., Ryan, J. C., & Vieira, P. (2017). Introduction. In M. Gagliano, J. Ryan & P. Vieira (Eds.), *The language of plants: Science, philosophy, literature* (pp. vii–xxxiv). Minneapolis: University of Minnesota Press. Retrieved 28 October 2020 from http://www.jstor.org/stable/10.5749/j.ctt1nxqpqk.3

Iovino, S., & Oppermann, S. (2014). Introduction: Stories come to matter. In S. Iovino & S. Oppermann (Eds.), *Material ecocriticism* (pp. 1–17). Bloomington and Indianapolis: Indiana University Press.

James, E. (2017). What the plant says: Plant narrators and the ecosocial imaginary. In M. Gagliano, J. Ryan & P. Vieira (Eds.), *The language of plants: Science, philosophy, literature* (pp. 253–272). Minneapolis: University of Minnesota Press. Retrieved 28 October 2020 from http://www.jstor.org/stable/10.5749/j.ctt1nxqpqk.16

Kirstinä, L. (2011). Vattumato olemisen suuressa ketjussa. Zacharias Topelius eläinten suojelijana. In M. Laakso, T. Lahtinen & P. Heikkilä–Halttunen (Eds.), *Tapion tarhoista turkistarhoille. Luonto suomalaisessa lasten – ja nuortenkirjallisuudessa* (pp. 55–62). Helsinki: SKS.

Kokko, Y. (1944). *Pessi ja Illusia*. Helsinki: WSOY.

Kokko, Y. (1966). *Alli. Jäänreunan lintu*. Helsinki: WSOY.

Laakso, M., Lahtinen, T., & Heikkilä–Halttunen, P. (2011). Johdatus lasten – ja nuorenkirjallisuuden luontoon. In M. Laakso, T. Lahtinen & P. Heikkilä–Halttunen (Eds.), *Tapion tarhoista turkistarhoille. Luonto suomalaisessa lasten – ja nuortenkirjallisuudessa* (pp. 9–28). Helsinki: SKS.

Laist, R. (2013.) Introduction. In R. Laist (Ed.) *Plants and literature: Essays in critical plant studies* (pp. 9–17). Amsterdam: Brill. Retrieved 18 October 2020 from ProQuest Ebook Central https://ebookcentral.proquest.com

Lassila, P. (1999). Kirjallisuus sodassa ja kulttuuritaistelussa. In P. Lassila (Ed.), *Suomen kirjallisuushistoria 3. Rintamakirjeistä tietoverkkoihin* (pp. 8–38). Helsinki: SKS.

Lloyd, M. (2016). Performativity and performance. In L. Disch & M. Hawkesworth (Eds.), *The Oxford handbook of feminist theory* (pp. 572–592). Oxford and New York: Oxford University Press. Retrieved 28 October 2020 from ProQuest Ebook Central https://ebookcentral.proquest.com

Martin, E. (1991). The egg and the sperm: How science has constructed a romance based on stereotypical male–female roles. *Signs*, 16(3), 485–501. Retrieved 31 May 2020, from www.jstor.org/stable/3174586

Melkas, K. (2011a). "On syntynyt uusi ihminen". Yrjö Kokon saturomaani Pessi ja Illusia eettisenä puheenvuorona luonnon puolesta. In M. Laakso, T. Lahtinen & P. Heikkilä–Halttunen (Eds.), *Tapion tarhoista turkistarhoille. Luonto suomalaisessa lasten – ja nuortenkirjallisuudessa* (pp. 128–137). Helsinki: SKS.

Melkas, K. (2011b). Arvoituksellinen luonto. Ihmisen ja luonnon oikukas rinnakkaiselo Anni Swanin sadussa"Aaltojen salaisuus". In M. Laakso, T. Lahtinen & P.

Heikkilä–Halttunen (Eds.) *Tapion tarhoista turkistarhoille. Luonto suomalaisessa lasten – ja nuortenkirjallisuudessa* (pp. 90–100). Helsinki: SKS.

Ryan, J. C. (2012). Passive flora? Reconsidering nature's agency through human–plant studies (HPS). *Societies*, 2(3), 101–121. Retrieved 28 October 2020 from doi:10.3390/soc2030101

Ryan, J. C. (2017). In the key of green?: The silent voices of plants in poetry. In M. Gagliano, J. Ryan & P. Vieira (Eds.), *The language of plants: Science, philosophy, literature* (pp. 273–296). Minneapolis: University of Minnesota Press. Retrieved 28 October 2020 from http://www.jstor.org/stable/10.5749/j.ctt1nxqpqk.17

Saukkoriipi, S. (2003). Pessi ja Illusia: uudistava sotasatu. In L. Huhtala, K. Grünn, I. Loivamaa, & M. Laukka (Eds.), *Pieni suuri maailma* (pp. 162–165). Helsinki: Tammi.

Seifert, L. C. (2015) Introduction: Queer(ing) fairy tales. *Marvels & Tales*, 29 (1), 15–20. Retrieved 28 October 2020 from www.jstor.org/stable/10.13110/marvelstales.29.1.0015

Simola, R. (2013). Yrjö Kokon Pessi ja Illusia. Oman aikansa lapsi—ja myös myöhempien aikojen. In A. Rastas (Ed.) *Kaikille lapsille. Lastenkirjallisuus liikkuvassa, monikulttuurisessa maailmassa* (pp. 174–196). Helsinki: SKS.

Swan, A. (1933). *Anni Swanin sadut*. Helsinki: WSOY.

Topelius, Z. (1866). *Läsning för barn. Andra boken.* Stockholm: Albert Bonniers Förlag. Retrieved 17 October 2020 from http://urn.fi/URN:NBN:fi–fd2017–00012381

Topelius, Z. (1867) *Läsning för barn. Tredje boken.* Stockholm: Albert Bonniers Förlag. Retrieved 17 October 2020 from http://urn.fi/URN:NBN:fi–fd2017–00012382

Topelius, Z. (1867) *Läsning för barn. Sjette boken.* Helsingfors: G. W. Edlund. Retrieved 17 October 2020 from http://urn.fi/URN:NBN:fi–fd2017–00012448

3 Aristotle on Plants

Life, Communion, and Wonder

Hallvard J. Fossheim

At its best, children's and YA literature fosters imagination in a way which helps the reader think anew about what one thought was familiar and trite. Plants in general and trees in particular are among the beings thus imagined and reimagined, allowing us to experience delightful and surprising potentials in our surroundings as well as in our own minds. These springs of wonder may be activated by children's and YA authors, but rarely are they entirely initiated there. An author like Tolkien, say, is able to awaken wonder by being steeped in a tradition rich in historical perspectives and alternative understandings of well-known things and events. Such openness crucially depends on and contributes to there being no watertight bulkheads between our intellectual pasts, our presents, and our imagined futures. It is all the more deplorable, then, if scholarly shortcomings cut us off from a main artery of insight celebrating plant life, communion, and wonder.

Accordingly, my aim with this chapter is to counterbalance a one-sided view of Aristotle as the nemesis of plants. Recent years have seen a welcome interest in the complexities of plant life in general and tree life in particular.[1] In its wake, there have been historical and philosophical investigations into how we came to think less of plants than we should. As invariably happens, such stories select their villains, and Aristotle has been appointed the arch enemy of plants in a now fairly broadly shared narrative. Among the central claims are that Aristotle is responsible for the exclusion of plants, that he thinks they have no proper end or structure, and (in the extreme) that he denies them a status as living beings.

To specify a little, Hall (2011), who generally provides a detailed and systematic treatment of relevant traditions, claims that Aristotle continues "the drive towards separation and discontinuity" (22) and that plant functions "are rendered as passive, mechanical processes" (23). Nealon (2016, 31–33) claims that Aristotle's plants grow without limit (misconstruing his sole textual reference, *De Anima* II.12.424a30–35, which is a completely unrelated passage on perception), and that they have no higher end. Also among the more sketchy approaches, Mancuso and Viola (2015) confuse Aristotles's endoxic method with assertion and take liberties with how to read those reports in concluding that Aristotle in *De Anima* claims plants lack soul and life (section "The Plant World according to Writers and Philosophers").[2]

DOI: 10.4324/9781032066356-5

Such contentions are double pronged. They involve both exegetical affirmations concerning what the Aristotelian corpus states, and related arguments about its *Wirkungsgeschichte*. Concentrating on the exegetical questions, I will try to argue that the criticism is at best one-sided, at worst absurdly mistaken. Aristotle as perhaps the most influential thinker in the Western tradition naturally has a lot to answer for—he seems to have taught us to think of the world in terms of things, to systematize rationality, to distinguish between practical, theoretical, and technical spheres, and so on. But today's tendency to denigrate Aristotle as botanist and to contrast him with Theophrastus, the Father of Botany, is not well founded.[3] Although the text *Peri Fytôn* which we today possess with his putative signature is without doubt a later work (by Nicolaus of Damascus), the ancient tradition lists a work of the same title by Aristotle, probably lost at an early date.[4] The extant work makes it clear that Aristotle had both knowledge of and interest in plants.

In what follows, I will show that Aristotle's hierarchy of vegetal, animal, and rational existence is not an exclusion of plants but a highlighting of their status as definitive of life. To the objector who replies that life is cheap in ancient thought, it will be demonstrated that plants are not just alive according to Aristotle, but exemplify completeness (in a way not available to, e.g. basic beings like grubs and certain insects). In fact, in us too it is the vegetative soul principle that ensures our participation in a crucial form of divinity. Building on a sketch of his teleology and metaphysics, we shall see that our relation to plants is ultimately not one of exclusion, but of communion and wonder.

The subject on which we now enter is highly complex. Considerations of space thankfully force me to simplify some matters and remain silent about others. While this means that a fuller treatment would also give more attention to the critical side of things (Aristotle does after all place plants at the basis of the global hierarchy of life, and he does say things about these forms of life that demand quite a bit of footwork to integrate into any account), it has seemed obvious to me that an effort to bring out a reasonable interpretation of some positive sides of Aristotelian botany is what is most urgently needed at this point in the debate. Ideally, we should also be able to use Aristotle as a living resource for reflection in the ongoing plant resurgence.

Aristotle's Teleology

In Aristotle's theory, plants are complete teleological beings. This also means that plants in Aristotle's book are ends in themselves, irrespective of their usefulness to us (which is also of course an end). In order to understand Aristotle's take on plants, then, one must understand his teleology. The word stems from *telos*, Greek for "end" or "aim". Describing and explaining human actions is typically done with reference to their end. I am grinding coffee beans in order to make myself a cup of coffee. Understanding the activity of grinding the beans requires seeing it in the light of the action as a whole, which will make sense of this part of the process. If enjoying a cup of coffee is the end, then this end is what explains earlier undertakings like grinding the beans and heating the water.

In all, Aristotle has an arsenal of four explanations or causes, that together cover everything you need to understand something: the material cause, the moving cause, the formal cause, and the teleological cause. While they interrelate in different ways depending on what is to be explained, and while they are all important, the teleological cause enjoys a certain primacy. This is because if we take away the end of things, they no longer make sense.

There is a tendency to think of Aristotelian teleology as requiring conscious intentionality. Nothing could be further from the truth. The coffee example typically does involve conscious intentionality. I may not have given it much thought—I could be acting on habit or deep in reflection about the paper I am writing—but some intentionality will typically be involved in settling that one is going to make a cup of coffee, and then carrying out the various parts of the project. This is not to say that is how all cases of teleology unfold, however. Aristotle does not consider consciousness an integral part of teleology.

On the contrary, thinking and planning is a way of doing something which is often better done without deliberate intent. The better craftsperson is not the one who is forced to stop and reflect during the process. In arguing for the absurdity of refusing purpose to nature simply because it cannot be seen deliberating, Aristotle informs us that "craft does not deliberate".[5] The amateur and learner typically will have their work fragmented by thinking, asking or doubting, where the properly trained expert just acts. The ideal is a structured and ordered set of movements leading irrevocably to a perfect result, without exception and without hesitation.

Aristotle distinguishes the craft from the person. Using literary production as our example, if you are asked to identify the maker of, say, a tragedy, while you are not utterly wrong in responding "Joanne", you are not being very helpful either. For a proper understanding of things, we need to see *as what* Joanne is producing the text. She is doing this not as just Joanne, nor as woman or as human being. The tragedy-producing capacity which, after much practice, is present in her is what is active and explains the production of the tragic play. This means that even pointing to her and answering "the craftsperson" is not the optimal reply. For a crafts*person* is also a series of other things. Aristotle thus distinguishes between the craftsperson and the craft in order to properly identify the moving cause in, say, the production of a tragedy. The craft is what is active and explains the production of the tragedy with a reference to the tragedy as end. Even if the craftsperson has to stop and ponder now and then, the craft itself does not.

Now let us take this necessarily pedantic insight over to a process present in nature: the production of the next generation. Aristotle enjoys summing up a basic principle from his embryology with the words "A human being makes a human being".[6] Aristotle's seemingly mundane dictum is crammed with meaning. The individuals belonging to each species characteristically bring forth new individuals that also belong to that same species. Elms produce elms, hawkweeds produce hawkweeds, rats produce rats, and humans produce humans. Now the process of generating a new living being is a paradigmatic example of Aristotelian teleology in action. The beauty of natural generation inheres in its orderly realization of a natural entity, and crucially, the core process is not a conscious or intentional endeavour.

Aristotle conceives of such agency as like the case of craft in one irreducible respect. Craft is best grasped as a set of complex, ordered movements. Similarly, natural generation is also a set of complex, ordered movements. Its material realization will differ from one group of living things to another. Among blooded animals, concerning which we have Aristotle's analysis preserved in writing, the material and medium for movement is the blood, while the physical agency in which the complex, ordered movements inhere is *pneuma* (vital heat or breath) consisting of hot air. This hot air thus holds the code for producing another like the parent and will under suitable circumstances do so. No consciousness is required (or even imaginable) for this process. Although the material bases might differ from one group to another, the general theory will be the same for all, including plants. In this sense, whether they lack the conscious awareness of themselves that some other beings possess does not make an iota of difference to their status as complete teleological beings.

Admittedly, beings differ quite a bit in how difficult it is to unravel their more specific goods or ends. In this respect, the plant is a tough customer. "The purpose [*to heneka tou*] resides in plants as well, although less articulated" (*Physics* II.8.199b10–12). This presumably means both that plants and their organs are not as easily grasped,[7] and that their lives are not as complex in articulation as those of the more advanced animals. Not admitting plants' teleological orderliness is nonetheless tantamount to doing away with nature; if this were the case (as claimed by people like Empedocles), anything could ensue from sowing a given kind of seed (*ibid.*, 199b13–14; cf. *Partibus Animalium* I.1.641b27–29).

To complete the picture of Aristotelian teleology, it should be borne in mind that craft is ultimately not the original for the model, but itself an appropriation gathered from nature. The original teleology is a natural one, and crafts enter the picture to insert themselves into these natural structures and ends. Cooking, for example, takes its cue from the digestion going on naturally in our stomachs, moving part of that process from the stomach to the pot heated by fire; the original end of digestion is given by our nature, and over time, a craft is developed to imitate and ease the work of nature. Another case in point is medicine, which inscribes itself into the massive, complex work our nature does to keep us healthy, to help the body back on track when this is necessary. The remedies of medicine enter the picture to assist nature, which still dictates the ends to which the craft is then calibrated. This also means, of course, that even to the extent that a craftsperson uses consciousness and intentionality as their means of doing the job, nature's original dictates and processes are not inherently conscious. Whether in human, wolf, or myrtle: teleological processes like digestion and growth are not at heart conscious or deliberative undertakings.

Energeia and the Metaphysics of Plants

In the field of metaphysics, Aristotle is known not only for his ontological categories and notion of substance, but also for fleshing out activity and ability as modes of being. This theory, of what in the literature is traditionally referred to as "actuality" and "potentiality", allows Aristotle to differentiate teleological developments in terms of being. For instance, as a human being, you have the ability to learn the craft of gardening. This is a first potentiality in relation to the activity of gardening. Once you've been through the training, you have obtained a second potentiality, namely, the ability to activate your gardening skills at any time without further processes of learning and developing; in Aristotle's parlance, this second potentiality is at the same time a first actuality. Once you are actually gardening, this is a second actuality or activity.

In this perspective, a soul is a first activity or actuality. That plants are part of this picture is made clear already by the choice of examples. "for we say that Hermes is in the stone, and the half in the line, and we call corn what is not yet ripe" (*Metaphysics* V.7.1017b6–8). Our use of language reveals that we speak of things according to potentiality and so acknowledge this mode of being even before we start philosophizing. The grain is acknowledged in the plant because the plant shares with craft a teleological structure which allows for *energeia.* [8] *Energeia* is Aristotle's technical term for the complete activity or actuality of something, whereas *dynamis* denotes the potential for such activity or actuality. Even in terms of time, priority goes to *energeia* over potentiality or *dynamis*:

> earlier in time than that which is already a man in *energeia*, and the grain, and the seeing, are the material and the seed and that which is able to see ... But before them in time, there were other things in *energeia*, out of which these came to be.
>
> (*Metaphysics* IX.8.1049b19–24)

In short, to the question which came first of the egg and the chicken, Aristotle's answer is the (mature) chicken. Similarly, plant being can only be grasped by appreciating the distinctive nature of the mature plant, of the relevant species, in its relation to the complete circle of generation.

Life and Soul

As I mentioned in this chapter's opening, critics have claimed that Aristotle thoroughly divides plants from animals (including humans), and even that he tends towards denying life to plants. Both claims are utterly mistaken. On the contrary, Aristotle argues for a deep connection between all living things, plants included, and he makes a big song and dance about the beauty and wonder of life (again, plants included).

Although it might sound odd, this is not to say that Aristotle does not distinguish hierarchically between humans, animals and plants. In order to see how these seemingly disparate approaches come together, we must address the Aristotelian notion of soul (*psychê*). A soul is the life principle of the being in question. Paying close attention to the parts and practices of animals—as does Aristotle—at once reveals that the same soul cannot be expected to control two species as different as, say, the fox and the swallow. Their organs are different, their forms of life are different, and so if soul is the controlling and uniting principle of each life, then it has little meaning to portray the same kind of soul as doing that job for both beings.

For Aristotle, then, *psychê* is the life principle of each being not in some abstract or vague way, but as what explains the being in question. Explaining a being means pointing to what makes it special or different from other beings, as well as what it shares with other beings. With this in mind, we turn to his hierarchy of souls. Its levels broadly conceived are those of the vegetative, the animal and the rational order. The distinction between the first two amount to a discernment between plants and animals. The distinction between the last two is, in the sublunary world, what sets human beings apart from all other beings.

Our Communion with Plants

Now there are at least three issues here. A first issue is whether the basic relation between plants and animals is ultimately a dichotomy or not (a methodological question). A second issue concerns the question of a hierarchy (a distinction in terms of worth). A third issue has to do with whether Aristotle is correct in denying *aisthêsis* (most often translated as "perception") to plants (an empirical question, albeit partly dependent on the question how to define *aisthêsis*). I will address each of these questions in turn.

As is his wont, Aristotle makes a fresh start on the question of soul after having gone through the theories of his predecessors.

We say then, assuming a fresh starting point for our inquiry, that that which has soul is distinguished from that which has not by living. But the word living is used in many senses, and we say that a thing lives if any one of the following is present in it—mind, sensation, movement, or rest in space, besides the movement implied in nutrition and decay or growth. Consequently all plants are considered to live, for they evidently have in themselves a capacity and first principle [*dynamin kai archên*] by means of which they exhibit both growth and decay in opposite directions, and in every direction, and they are nourished and continue to live, as long as they are able to absorb food. This capacity to absorb food may exist apart from all other powers, but the others cannot exist apart from this in mortal beings. This is evident in the case of plants; for they have no other capacity of soul.[9]

It is crucial to see that the distinctions between plants, animals, and humans do not establish divides, but a communion. This is perhaps most easily seen in the relation between animals and the human (rational) being. There is no divide between human and animal, for the simple reason that humans are animals—thus the locution "rational animal". The same applies to the differentiation between plants and animals. The capacity for nourishment and growth[10] is most strongly and obviously displayed in plants, where it constitutes their defining activity. No living being, however, can exist without this power. An animal soul is also a plant soul. Putting the matter paradoxically, being an animal partly means being a plant by other means. Being an animal involves performing some of the plant functions in other ways, determined and made possible by the combined abilities of perception and movement. Similarly, being human centrally means doing animal stuff in a specifically human, that is, rational, way. So while perception and reason in a way change everything, in another way they do not. They are not added features that can simply be tacked onto an existing life form, but abilities that somehow transform while holding on to what else is there. Successful living is an existence that manages to integrate the various capacities of soul into a harmonious whole.

We are all—as animals and as humans—at the same time plants, in the technical sense that we all possess plant souls as integrated into our souls. Aristotle also expresses this by saying that the nutritive faculty (*to threptikon*) is present in potentiality in the next in the series, that is, the sensitive faculty (*to aisthêtikon*).[11] This is why he insists that an inquiry into soul must not only speak generally, but examine each kind of being: in order to understand plant souls, there is no substitute for actually investigating plants.[12] This insight forms the very basis of Aristotle's restructuring of psychological as well as biological science. At the same time, there is no denying that he takes the later functions to be higher.[13] The Aristotelian soul hierarchy has no qualms that in the big picture, animals are higher than plants, humans than the other animals, and gods than human beings. What is most widely shared is less exalted. In the case of reason especially, *logos* is divine in the specific sense that the god or gods have this as their unique characteristic, which means that reason defines the highest being in the universe (which is, then, not the human being). Behind these two more concrete reasons for emphasizing the higher status of animals and humans when compared to plants, there is also the fact that Aristotle takes everything to be imbued with worth. This is also a necessary feature of his teleology, in that something can be explained by showing that it is better thus. Normativity in this sense invades even his understanding of directions in the cosmos (up is better than down, right is better than left, and front is better than back).[14]

The *Aisthêsis* Question

A question one might still pose is why Aristotle denies *aisthêsis* to plants.[15] The above analysis of teleology hopefully provides most of the answer. Aristotle's teleology is simply not so banal that it reduces complex goal-directedness to sensory awareness. At the same time, Aristotle's teleology builds understanding on function. Some functions require sensory awareness. In these processes, the speed and local motions required are probably teleologically central to explaining a need for sensation. *Aisthêsis* is then first and foremost a particular way in which to perform another function (nourishment or procreation), where that latter function is logically and physically independent of sensation.

It is of course important in this connection to distinguish between what we today happen to accept as sensible opinions and what represents the best empirical information available in the historical context we investigate.[16] The ancient Greeks did not know the function of leaves (photosynthesis is a fairly fresh theory), and certainly were not familiar with empirically founded theories of tree communication that have gained a foothold only in the last generation or so. Given that what one has to build on are close empirical observations of the sort provided in Aristotle's work and in Theophrastus' *On Plants* and *Causes of Plants*, combined with a sophisticated teleological explanatory model, postulating *aisthêsis* in plants could reasonably be deemed both unfounded and unnecessary.[17]

The question about plants and *aisthêsis* has two issues built into it, however. Not only the issue of how plants live and interact with their environment, but also the issue of how Aristotle construes *aisthêsis* is relevant to providing a full answer to the original question. Let me at this point briefly venture two rather speculative thoughts, on which none of what I have said so far hinges. I do not present them as final answers, but as possible venues for further queries.

The first thought is that in his analysis of *aisthêsis*, Aristotle reaches the conclusion that touch is the most basic among the five senses.[18] This means that touch has a position among the senses analogous to the position of plants among living things, of being necessarily present in the widest genus. If a mortal entity possesses *aisthêsis*, then eo ipso it possesses the sense of touch. Could it be that building from this as a premise, Aristotle also takes plants to be cut off from being attributed with *aisthêsis* partly because they do not do not generally present us with any sign that they respond to touch?[19]

The second thought concerns the relation between *aisthêsis* and movement. While perception and not movement is what settles the differentiation between animals and plants, it can be helpful, in order to grasp what Aristotle means by "perception" (*aisthêsis*), to keep in mind that he normally takes the two to go together. According to his teleology, nature would be acting in vain if it gave *aisthêsis* to a being that was incapable of moving. This may be why Aristotle in the *Historia Animalium* pauses to discuss, and report non-committally, sponges' putative perception: What is the point for a sponge of perceiving anything, when it is incapable of running, slithering, swimming, or flying away from or towards the object? In this case, the sponge's contended tightening motion provides a possible answer.[20] We can presume, then, that *aisthêsis* in the paradigmatic case is intimately connected with being able to avoid or pursue the object of perception by moving from one place to another.

Both these thoughts share as a point of departure a caution regarding exactly what Aristotle means by *aisthêsis*. We should not take for granted that it means whatever we take "sensation" or "perception" to mean in our everyday or professional talk.

The Wonder of Plants

Plants are complete beings, which is to say that they too strive for divine immortality through reproduction. By now we know that "strive" does not imply planning or intentionality; that would anyway not be directly applicable, since Aristotle's meaning is not that even animals consciously strive for immortality. Animals strive for things like food and sex, each in their characteristic manner. It is only the natural philosopher who realizes that such strivings are nature's way of emulating the eternal life of the divine, as mortal individuals partaking in an eternal chain of being. It is the plant soul—in other species present as the nutritive faculty broadly conceived—to which biological participation in eternity, and thus divinity, belongs. This too is ultimately part of the wonder of plant life, available as a source of amazement for whoever is willing to pay attention to our surrounding world.

As part of his opening methodological remarks in the *Partibus Animalium*, Aristotle makes a broad distinction between two kinds of objects for science, "the ungenerated and indestructible things, and the things that partake in generation and destruction" (I.5.644b23–25). While the eternal is obviously higher and more divine, the living entities around us have one great advantage as sources of insight. "Concerning the destructible plants and animals, we have a cognitive advantage in living among them" (ibid., 644b28–30). For the philosophical person, the study of even aesthetically less agreeable living things provides immeasurable joy (ibid., 645a8–11).[21]

"Therefore, we must not childishly react with disgust in investigating the more graceless living things [*zô(i)ôn*]. In each and every one there is by nature something wonderful" (ibid., 645a15–19). We can safely enter such investigations knowing there is something of nature and of beauty (*tinos physikou kai kalou*, ibid., 645a23) in every item. The references to the philosophical person, to wonder and to beauty in these passages are not accidental but intimately related. Philosophy starts with wonder, and beauty to Aristotle is a question of function and purpose.

Part of the wonder of plants is that they are complete in the sense mentioned above: they are beings that participate in the circle of generation, with individuals producing individuals of the same kind. This is shared with animals, including ourselves; but it is as vegetative beings that we are wondrous in this way. The plant soul houses our part in generational eternity. Thus, pulling these strings together, the nutritive soul belongs to all the other living creatures besides man, and is the first and most widely shared faculty of the soul, in virtue of which they all have life. Its functions are reproduction and the assimilation of food. For this is the most natural of all functions among living creatures, provided that they are perfect and not maimed, and do not have spontaneous generation: viz., to reproduce one's kind, an animal producing an animal, and a plant a plant, in order that they may have a share in the immortal and divine in the only way they can; for every creature strives for this, and for the sake of this performs all its natural functions (*DA* II.4.415a24-b2).

Plants not only partake in the eternal, complete and divine, but are its basic vehicles. Furthermore, according to this passage, even specifically animal functions are ultimately directed at fulfilling this plant function of divine immortality.

How do plants do it? As usual, Empedocles takes the brunt of Aristotle's ire, for not being able to see that his theory did not even begin to uncover how it is that a plant is not just an haphazard amalgam of natural bits, but a *unity*. In an Empedoclean world, the fiery parts of plants would drift upwards, while the earthy parts would sink downwards. The challenge, then, is to begin to see why this is not the case. An aspect of the wonder of plants is thus that they are not simply an element, such as "earth" or "earthy" (although they are that too), but complete organisms controlling and shaping earth or the earthy. There is a *physis* generating unity and constituting life—"what is it that holds fire and earth together when they tend to move in contrary directions?" (*DA* II.4.416a6–8)

One way in which to phrase the answer, then, is that a plant, like an animal, has a nature. A nature or *physis* is an internal principle of movement and standstill[22]—in the plant's case, as the individual plant does not characteristically possess local movement from one place to another, the movements of growth and the stability of upholding itself. The integrity of such a natural being sets it radically apart from the products of craft, which only have an external such principle (the craftsperson who first produced, and then repairs, the bed or other implement).

Connecting this with our earlier *energeia* discussion, the plant soul partakes in *energeia* as activity in its relation to food. Again, the simplistic, Empedoclean approach would be to claim that fire is what performs digestion. But fire has no proper measure in it, as anyone knows who has witnessed it devouring a house or a forest. Fire

> is in a sense a contributory cause, but not absolutely the cause, which is much more properly the soul; for the growth of fire is without limit, so long as there is something to be burned, but of all things naturally

> composed there is a limit or proportion of size and growth; this is due to the soul, not to fire, and to the essential formula rather than to matter.
> (*DA* II.4.416a15–19)

A plant's nature is its soul, co-defining life for all earthly beings and providing active participation in divine eternity as part of the package.

The Presence of Plants

Finally, how can we best characterize the presence that plants do have? As we have seen, Aristotle denies *aisthêsis* to plants, thus provoking exegetical questions about the nature of *aisthêsis* in our traditions. This is what we might call the Aristotelian pole of the question. The vegetational (dare I say fruitful?) question pole, by contrast, concerns how we might most truthfully sketch the perception-like abilities that plants seem to have. Here, Aristotle's critical reflections on both *aisthêsis* and methodology can hopefully help us move beyond the more agenda-like attitudes that use history primarily for setting up straw men (pun not intended).

Aristotle has perhaps a suggestion for how to work towards more adequate appreciations of the presence that plants have. For some things, we have a settled understanding, with full definitions and crisp borders. At the other extreme, there are phenomena which we see only vaguely and without developed conceptual tools. For such predicaments, whether they are due to shortcomings in us or to a manifold in the phenomenon, Aristotle recommends applying analogy as an approach.[23] We saw this exemplified above, in his treatment of *energeia*. The Hermes, the half, and the unripe corn: these are not just three examples, to be unified under a universal concept, but analogies.

Aristotle suggests that where we have almost nothing to go on by way of direct experience, analogy constitutes a resource. Concerning the presence of trees, then, Tolkien's Ents or other fantastical plants (or plantish beings) in children's literature may be as important a source in nourishing the required imagination and insight as, say, an empirical study or a philosophical analysis.

Notes

1 In a spate of publications, an outstanding bestseller remains Wohlleben (2016).
2 In what follows, I will not refer to individual instances of what I take to be misreadings of Aristotle. This is primarily because my aim is to provide a reasonable interpretation of Aristotle, not to file away at what I consider to be less sensitive readings of him.
3 Theophrastus, acknowledged as a major contributor in the history of botany, was Aristotle's favourite pupil and close co-worker in biology, to whom Aristotle also entrusted the execution of his testament. Egerton (2001) sounds a healthy note of caution concerning the biological works: "Later, the zoological treatises were attributed to Aristotle and the botanical treatises to Theophrastus, even

though they are similar in organization, style, and outlook" (p. 149). For an informative and enjoyable account of Aristotle's career which also includes intriguing detective work on the role of Theophrastus in settling Aristotle's estate and heritage, cf. Natali (2013).

4 Drossaart Lulofs (1957) suggests that remnants of it may be found in the later compilation by Nicolaus.

5 *Physics* II.8.199b27–28. Generally, translations from the *Physics, Metaphysics*, and *Partibus Animalium* are my own.

6 e.g. *Physics* I.1.193b8–11 and III.2.202a11–12; cf. also *Physics* VIII.5.257b10.

7 That plant organs and functions are not easily discerned does not mean they are not there: Aristotle assures us (with examples) that plants have organs even though they are simple (*DA* II.1.412b5–9).

8 Aristotle in fact seems to subscribe to a double teleology for all species of plant and animal: one building from the *energeia* of the beings themselves, and another, holistic one referring everything else to the good of the human being. An excellent presentation, with further references, is Sedley (2007). The ambiguity might be present in the grain example; as Marder (2014, esp. 22–23) points out, *sitos* has meanings spanning from wheat, over grain-qua-food, to staple food taken generally.

9 *De Anima* (hereafter *DA*) II.2.413a24-b2. Citations from *DA* are from W. S. Hett's Loeb translation.

10 Reproduction is proper to this same capacity in the case of complete living things because reproduction is a substantial (as opposed to merely quantitative) growth-like process, with motions transcending the individual.

11 *DA* II.3.414b29–415a3.

12 Ibid.

13 E.g. *De Generatione Animalium* I.23.731a25–b4.

14 Cf., e.g., *On the Heavens* II.5.288a3–12.

15 Repeatedly; e.g., *DA* I.5.411b27–30 and II.3.414a32–b1. This is a complex question, and I only bring up two considerations. Kent Sprague (1991), interestingly elaborates on how their earthiness relate to plants lacking *aisthêsis*. Coming from a different angle, Bos (2010; cf., e.g., 832–835) also sketches the implications of earthiness in animals.

16 An excellent sketch of this methodological ideal is Frede (1988).

17 Cf., e.g., *Partibus Animalium* I.5.644b26–31.

18 *DA* II.2.414a3–4, II.3.414b3–5.

19 *DA* II.12.424a33-b3. That plants do not perceive means that they have no mean (*mesotês*) for receiving the form without the matter. Receiving form without or with matter is the difference between seeing red (the form without the matter) and being made red (form with matter). The *DA* passage reminds us that plants are physically affected by heat and cold (seasons and weather).

20 *Historia Animalium* V.16.548b10–15.

21 In this passage, the context makes it clear that plants too should be taken as objects of the general characterizations (also, plants as well as animals are by nature more our own—*tês fyseôs oikeiotera*), although the programme statement limits Aristotle's ensuing investigation to animals. Skemp (1947, 56) is of course right that Aristotle also makes much of distinguishing between plants and animals, thus dividing *zônta* into *zô(i)a* and *fyta*, but what matters here is which capacities and characteristics living things are allotted. Cf again Skemp's main object of study, Plato's *Timaeus*, who states that plants are *zô(i)a* because they *zên* (live) (77B). Carpenter (2010) has suggested that this fine-tuned position in Plato's *Timaeus* between the recognizably present and the merely existing is intimately connected to the cut-off point for soul transmigration: plants' third rank *aisthêsis* lacks the potential for an integrated view of the whole and the

good which characterizes all beings involved in transmigration. Plants simply go on doing their thing, and in this also resemble rivers and mountains, their only goodness the fact that they hold bottom positions in a whole that is good. This does not abolish the famous statement that humans are plants with our heads (that is, our roots) in the heavens (*Timaeus* 90A-B).

22 *Physics* II.1.192b8–20.

23 For a treatment of Aristotle's uses of analogy, see Lloyd (1996).

References

Aristotle. (1924). *Metaphysics*. A revised text with introduction and commentary by W. D. Ross. Oxford: Clarendon Press.

Aristotle. (1936). *Physics*. A revised text with introduction and commentary by W. D. Ross. Oxford: Clarendon Press.

Aristotle. (1957). *On the Soul (De anima)*. Cambridge, MA: Loeb Classical Library. Translated by W. S. Hett.

Aristotle. (1961). *Parts of animals, movement of animals, progression of animals*. Cambridge, MA: Loeb Classical Library. With an English translation by A. L. Peck [Parts of Animals]. With an English translation by E. S. Forster [Movement of Animals, Progression of Animals].

Aristotle. (2000). *On the heavens*. Cambridge, MA: Loeb Classical Library. With an English translation by W. K. C. Guthrie.

Bos, A. P. (2010). Aristotle on the differences between plants, animals, and human beings and on the elements as instruments of the soul (De Anima 2.4.415b18). *The Review of Metaphysics*, 63(4), 821–841.

Carpenter, A. D. (2010). Embodied intelligent (?) souls: Plants in Plato's *Timaeus*. *Phronesis*, 55, 281–303.

Drossaart Lulofs, H. J. (1957). Aristotle's *Peri Fytôn*. *The Journal of Hellenic Studies*, 77(1), 75–80.

Egerton, F. N. (2001). A history of the ecological sciences, part 2: Aristotle and Theophrastus. *Bulletin of the Ecological Society of America*, 82(2), 149–152.

Frede, M. (1988). The history of philosophy as a discipline. *The Journal of Philosophy*, 85(11), 666–672.

Hall, M. (2011). *Plants as persons: A philosophical botany*. Albany, NY: SUNY.

Kent Sprague, R. (1991). Plants as Aristotelian substances. *Illinois Classical Studies*, 16(1–2), 221–229.

Lloyd, G. E. R. (1996). The unity of analogy. In Lloyd, *Aristotelian explorations* (pp. 138–159). Cambridge: Cambridge University Press.

Mancuso, S., & Viola, A. (2015). *Brilliant green: The surprising history and science of plant intelligence*. Washington: Island Press. [Translated from the Italian original 2013 edition by Joan Benham.]

Marder, M. (2014). *The philosopher's plant: An intellectual herbarium*. New York: Columbia University Press.

Natali, C. (2013). *Aristotle: His life and school*. Princeton: Princeton University Press. [Orig. published 1990.]

Nealon, J. T. (2016). *Plant theory: Biopower & vegetable life*. Stanford, CA: Stanford University Press.

Saguaro, S., & Thacker, D. C. (2013). Tolkien and trees. In P. Hunt (ed.), *New casebooks: J. R. R. Tolkien* (pp. 138–154). Basingstoke: Palgrave Macmillan.

Sedley, D. (2007). *Creationism and its critics in antiquity*. Berkeley, CA: University of California Press.

Skemp, J. B. (1947). *Plants in Plato's Timaeus. The Classical Quarterly*, 51(1–2), 53–60.

Wohlleben, P. (2016). *The hidden life of trees: What they feel, how they communicate*. Vancouver: David Suzuki Institute/Greystone Books. [Translated from the German original 2015 edition by Jane Billinghurst.]

Part II

Plants in Folklore and Fantasy

4 Come into the Garden, Alice

Rude Flowers, Dream-Rushes, Aphasic Woods, and Other Plants in Lewis Carroll's Nonsense Worlds[1]

Francesca Arnavas

Introduction

This chapter explores the multifaceted significance of plants and vegetation in Lewis Carroll's *Alice* books, *Alice's Adventures in Wonderland* and *Through the Looking-Glass and What Alice Found There*. If plant-related imagery has shaped much of Western culture, as recently outlined by Randy Laist (see his edited volume *Plants and Literature: Essays in Critical Plant Studies*), the Victorians in particular elaborated and extended in their literary and artistic endeavours the reciprocal symbolic exchange between plants and culture. Carroll took this link even further, by using it as a way to parody and question contemporary common beliefs. To better capture this perspective, adopting a critical focus on how plants are depicted offers a new way of interpreting and decoding Carroll's nonsense masterpieces. The role played by gardens, flowers, and woods in the *Alice* books is an active one: vegetation does not only offer a setting for Alice's vicissitudes, but it also influences her actions and modifies her identity. Plants even become characters in the story (such as the talking flowers and the snap-dragonfly, a hybrid between an insect and a plant). Despite a general recognition of the importance of these plant-related elements in the *Alices* (T. S. Eliot, for instance, refers to Alice's garden in the beginning of his poem *Burnt Norton*, 1943), a thorough study of how this perspective can encourage new readings of Lewis Carroll's masterpieces has not yet been offered.[2] This chapter provides a specific focus on the role of plants in the *Alices*, reflecting on their complex narrative function by connecting it to the motif of the Victorian garden, to the enchanted forest in the children's literature tradition, and to Victorian literary tropes of flowers and plants. This chapter argues that in each case—garden, flowers, and woods—Carroll, through the use of inversions, parodies, and by frustrating the readers' expectations, offers innovative ways of questioning beliefs and traditions. Victorian and fairy tale tropes connected to vegetation are transformed in order to highlight their link to rigid and oppressive convictions. The apparent nonsense of the Carrollian mad gardens, rude flowers, and aphasic woods actually performs as a key device to discuss problematic assumptions and undermine Victorian

DOI: 10.4324/9781032066356-7

Figure 4.1 Sir John Tenniel, *Alice in the Garden of Live Flowers*, 1871. Illustration for Lewis Carroll's *Through the Looking-Glass and What Alice Found There.*

traditional certainties, introducing new perspectives, such as an anticonventional conception of the girl, and posthumanistic viewpoints. Zoe Jaques claims that "by imagining 'being' as operating beyond bodily and environmental constraints, children's literature can offer sophisticated interventions into debates about what it means to be human or non-human and offer ethical imaginings of a posthuman world" (2015, p. 5). Carroll's *Alice* books function as a particularly relevant case in point, as stressed by Jaques herself (2015, pp. 23–71), with their radical overthrowing of fixed categories.

Gardens and Reversals: Childhood Eden as the Reign of Chaos

The Victorian era witnessed a veritable obsession with gardens and flowers, and, in particular, with the association between women and gardens: "what Michael Waters has called, in Ruskinian language, 'the Queen of the Garden role' in Victorian culture" (Bilston, 2008, p. 2). Victorian gardens were conceived as "a sort of heaven on earth, an ideal world in which the promises of balance and order are fulfilled" (Bilston, 2008, p. 5). Even more prominent was the connection between girls and gardens, which culminated in Frances Hodgson Burnett's *The Secret Garden*, in 1911. But even before Mary Lennox, Victorian girls were often depicted in relation to gardens, whether in association with gender-related reflections, matters of education and proper behaviour, or celebrations of purity and grace. For instance, Elizabeth Watts wrote in *Flowers and Flower Gardens* (1866) how taking care of the garden space could help in perfecting and developing womanly qualities in little girls. In Juliana Horatia Ewing's *Mrs. Otherway's Remembrances* (1869), the wandering of a girl named Ida in the garden perpetuates the celebration of childhood as a sort of Arcadian state of perfection. Another possible example is George MacDonald, who in his fantasy worlds often describes girls in the setting of enchanting (and sometimes enchanted) gardens, surrounded by snowdrops, bluebells, roses, daisies, and often fairy-flowers as well (*Phantastes*, 1858, *The Day Boy and The Night Girl*, 1880, *The Princess and the Goblins*, 1872, *Heather and Snow*, 1893). Among Victorian writers then, there was, as Humphrey Carpenter points out, "a fondness for the symbol of a garden or Enchanted Place, in which all shall be well once more" (1985, p. 9).[3]

Lewis Carroll's *Alice* books take this traditional Victorian children's literature motif, and turn it upside down, as they do with many other typical tropes of children's literature and moral messages. The Wonderland garden is the celebration of nonsensical madness, and the Looking-Glass garden acts as a similar narrative device, upsetting Alice's and Victorian readers' assumptions about how a proper Victorian garden should be. The Victorian conception of the garden as a safe space for girls discloses in Alice's gardens its inherent dark side: it becomes a claustrophobic space of danger, governed by senseless rules. This reversal of conventional representations functions also as a way to highlight the problematic aspects of those representations themselves.

The important and disclosing role assigned to gardens is stressed already at the very beginning of the first *Alice* book: "she knelt down and looked along the passage into the loveliest garden you ever saw. How she longed to get out of that dark hall, and wander about among those beds of bright flowers and those cool fountains" (*AAIW*, pp. 15–16). This quotation appears immediately after Alice has jumped down the rabbit hole and reached its bottom, at the very start of her adventures. The sight of the garden is one of the main triggers for Alice's adventures. It is in order to get through the narrow doorway to reach the garden that she starts eating and drinking strange potions to shrink or grow. A similar scenario occurs at the beginning of the second *Alice* book, *Through the Looking-Glass and What Alice Found There*: after having gone through the bright silvery mist of the mirror, Alice finds herself in the Looking-Glass House, and her plan of action is the following: "let's have a look at the garden first!" (*TTLG*, p. 156). It is to be able to get there that she starts understanding how the Looking-Glass world works (by going backwards in order to go forwards). The desire to come closer to a garden is thus behind Alice's experiences in both the fantastic worlds of Wonderland and Looking-Glass Land: the garden is the focal point of attention which causes Alice to go ahead with her adventures.

Let us consider the two gardens separately and in more detail. Wonderland's garden encourages Alice's more conscious actions. In general, Alice's behaviour in Wonderland is marked by erratic, irrational decisions, which do not seem to follow any clear pattern; however, the aim of getting to the garden is the only distinct purpose which guides Alice in the chaotic Wonderland universe:

> "'the first thing I've got to do,' said Alice to herself, as she wandered about in the wood, 'is to grow my right size again; and the second thing is to find my way into that lovely garden. I think that will be the best plan".
>
> (*AAIW*, p. 45)

Later on, "the next thing is, to get into that beautiful garden—how *is* that to be done, I wonder?" (*AAIW*, p. 58).[4] As is well known, Alice's experience in Wonderland is a product of her dreaming mind: "the whole thing is a dream" (Cohen, 1989, p. 29), wrote Carroll himself, and what he achieves in his narrative masterpieces is a dream-like writing, in which characters and landscapes switch and change in a fluid flow with no apparent meaning or logic. "Writing like a dream is exactly what Carroll attempted to do in this story" (Douglas-Fairhurst, 2015, p. 125), and in dreams goals keep changing and get forgotten and replaced easily. Nevertheless, the aim of coming nearer to the garden seems to be an invariable constant in Alice's moves through the land of her dreams, and it appears somehow to give some narrative structure to an otherwise orderless fictional scenario.[5]

However, when "she found herself at last in the beautiful garden, among the bright flower-beds and the cool fountains" (*AAIW*, p. 81), things are not as expected. The garden is revealed to be the Queen of Hearts' mad-croquet ground, which is the triumph of Wonderland's chaos and folly. It is the place in which all boundaries are lost. Even the basic distinction between animate and inanimate does not stand anymore: animals like flamingos and hedgehogs are treated like objects (they are respectively the mallets and the balls of the croquet game), while what are normally considered inanimate beings like cards are actually the players—the Queen, the King and their subordinates. As Hugh Haughton puts it, "the Wonderland garden is no childhood Eden, but a life-and-death croquet match presided over by a homicidal Queen shouting 'off with their heads' every second minute" (2003, p. xliii). This parodic reversal, using the garden as a site of questioning traditional beliefs, reminds one of Christina Rossetti's use of fruit in *Goblin Market* (1862), as remarked by Akemi Yoshida (2013): Rossetti takes the way in which contemporary Victorian artists and writers represented the forbidden fruit in relation to women, and questions its assumptions in her bizarre poem/fairy tale. Similarly, the correspondence between idyllic gardens and perfect angelic girls is overthrown and parodied in the Wonderland garden: plants, for both Rossetti and Carroll, work as means for undermining Victorian conventional convictions, especially those connected with feminine identity.

Alice herself has undergone a number of transformations, having changed size and having been mistaken for different beings, but perhaps the most significant of her multiple Wonderland identities is in this case that of the serpent. The pigeon takes her for a serpent wishing to eat her eggs, and Alice cannot help but admitting that "little girls eat eggs quite as much as serpents do, you know"—to which the pigeon logically replies "if they do, why, then, they're a kind of serpent" (*AAIW*, p. 57). In the Wonderland garden, the little girl is thus the predatory serpent. If for Rossetti, Eve's temptation and fall is re-written and parodied in Laura's eating of the Goblin's fruit, Carroll goes even further by linking innocence and temptation directly, associating the girl with the serpent itself. It is clear now, then, how the narrative function of the garden as a structural and orderly one is just an initial misleading clue: Carroll uses it to reinforce instead the celebration of the topsy-turvy pattern of his dream-like narration, reversing in this way the expectations one could have formed about a typical Victorian garden, with a pretty little girl happily pottering around the flower-beds. The serpent-Alice here is using animals as objects in a game, and is in constant danger of being beheaded.

In the second *Alice* book, *Through the Looking-Glass and What Alice Found There*, the configuration of the dream-world follows more precise rules (the space is represented as a chessboard) and Alice has a specific goal, going from being a pawn in the chess game to becoming a queen. Despite this, the first thing Alice wants to do after she has gone through the mirror is have a look at the garden, which is not then her final aim, as in Wonderland, but her *first* one. It is precisely in the garden that she meets the Red Queen and realizes she is a pawn and she has to move on the chessboard in order to become a queen herself (*TTLG*, pp. 171–176). In the beautiful flowery enclosed space the girl is just a pawn in a pointless game: Carroll poignantly satirizes how girls were portrayed in Victorian culture, and emphasizes how constraining and rigidly encaging this portrayal was.[6]

If Wonderland's mad-croquet ground inverts and displaces common presumptions, the Looking-Glass garden takes the inversion *literally*, since it is located on the other side of the mirror, where "things go the other way"

(*TTLG*, p. 147). Even just walking there from inside the house is not as easy, since in the Looking-Glass world you have to go backwards in order to proceed forwards. The garden itself is no Victorian Edenic place: there is indeed "a large flower-bed, with a border of daisies, and a willow-tree growing in the middle" (*TTLG*, p. 166), but once again there is nothing idyllic and particularly pleasant about it; the flowers can speak, and they are rude and aggressive towards Alice, the Red Queen Alice meets is bossy and makes Alice run to keep still, and quench her thirst with dry biscuits, and, as said, Alice finds out that she is a mere pawn in a game of chess. In this sense, as in *Alice's Adventures in Wonderland*, the garden plays an important role in the process of demystification and parody that Carroll enacts in his fictional works, proving the instability of conventional beliefs and explicitly ridiculing the Victorian literary trope of the girl in the flower-garden.

The Language of Flowers

Flowers and Violence: Turning the White Roses Red

If gardens—and young girls in them—held a special place in the Victorian imagination, it is not an overstatement to say that the Victorian era was *the* flower age: "that Victorians loved flowers was obvious" (Seaton, 1985, p. 255). This passion for botany was reflected in the proliferation of gardening amateurs, the publication of a huge volume of books about gardening, a rising interest in flower arrangements and flower compositions, and a widespread presence of flower references in poetry, novels, artworks (see Seaton, 1995). Seaton also points out the "Victorian preoccupation with flowers as primary expressive elements in human experience" (1985, p. 255) which also found a literary elaboration in works by writers such as George Eliot, Elizabeth Gaskell, and Charlotte Brontë (Engelhardt, 2013). It was, then, in this specific flower-obsessed cultural context that Lewis Carroll created his "garden of live flowers" in the Looking-Glass Land. Gillian Beer points out that "Carroll had a considerable interest in botany and read, and in many cases owned, a number of contemporary books on wild flowers, trees, and plants" (2016, p. 150). As for the gardens, applying to Carroll's representation of flowers the critical lens of plant studies helps in providing a new interpretative tool to enhance the significance of the *Alice* books' critique of Victorian culture. Since "plants and human culture are so deeply interwoven" (Laist, 2013, p. 14), the importance of the innovative depiction of flowers in the *Alices*, during the Victorian flower age, cannot be underestimated.

Several critics (see Green, 1971; Gardner, 2001) have recognized that the talking flowers chapter is inspired by a few passages in Tennyson's *Maud* (1855), the notorious part beginning with the verse "come into the garden, Maud" (pt. 1, section XXII). If in *Maud* the language of flowers, so popular at the time, is used to portray the anxious waiting of a lover for his lady, with the different flowers expressing his own passionate feelings, in Carroll's garden the flowers do not need to symbolize anything, since they are able to speak themselves. What they say, however, is not what they are supposed to represent: if lilies meant purity and innocence (as in Tennyson's poem), the Tiger Lily here has a loud and aggressive temper, addressing Alice in a very impolite manner (*TTLG*, pp. 166–168). The Violet, linked to modesty, as in the comparison with Maud's eyes (v. 891), here says about Alice "I never saw anybody who looked stupider" (*TTLG*, p. 168). The flower whose presence is most prominent in *Maud* is the rose, linked to the speaker's intense emotions, to Maud's beauty and to their love. In the Looking-Glass garden, the Rose also has the most relevant role, being the one who talks to Alice most, and who tells her about the Red Queen's presence in the garden. Nevertheless, the messages the Rose delivers to Alice are not what one may expect: she starts the conversation by saying about Alice "her face has got *some* sense in it, though it's not a clever one!" (*TTLG*, p. 166). Afterwards, she remarks to her "it's *my* opinion that you never think *at all*" (*TTLG*, p. 168), and her final observation to Alice is "you're beginning to fade, you know—and then one can't help one's petals getting a little untidy" (*TTLG*, p. 169).

The radical conceptual inversion Carroll applies in his *Alices* finds in the representation of the live flowers one of its most significant and extreme examples: the flowers turn from being delicate and splendid carriers of important messages to rude and insulting creatures, overturning in this way all the Victorian assumptions about the nature and language of flowers. Robert Tyas's *The Language of Flowers: or, Floral Emblems of Thoughts, Feelings and Sentiments*, published in 1869 (*Through the Looking-Glass* came out in 1871) precisely assigns to each flower, and to each different colour a single flower can have, a very specific meaning. Carroll was notoriously very sensitive to non-human creatures' rights (as he clearly shows in his treatise against vivisection "Some Popular Fallacies about Vivisection"): with his garden of live flowers he explicitly assumes what nowadays would be labelled as a "human-plant studies viewpoint", "a broad framework for re-evaluating plants, their representations, and human-plant interaction" (Gagliano et al., 2017, p. xvii). The garden is a place where "meaning is entirely occupied by one species, who interpret everything in their own terms" (Beer, 2016, p. 149), in a clear parodical reversal. Carroll's flowers interpret humans with flower-like parameters, ironically mimicking human behaviour towards plants. As critics have already remarked, Carroll's *Alice* books repeatedly show the influence of Darwin's works, with their references to evolution, taxonomy, extinction. Darwin was also "a distinguished botanical experimentalist, who departed from the classical view of plants as insensate to reveal their sensitivity and power of movement, akin to that observed in some animals" (Gagliano et al., 2017, p. xii), and it can be argued that Carroll, being well aware of Darwin's studies, could have been influenced by this perspective while describing his live flowers.

As seen, this reversal of meaning also includes parody of well-recognized literary works, such as Tennyson's *Maud.* It is also possible to trace a comparison with the poem *The Sensitive Plant*, written by P. B. Shelley in 1820, in which a delightful garden is described as an "undefiled Paradise" (v. 58), with roses, violets, lilies and daisies, and a "ruling Grace" (v. 116), a radiant young lady attending to the plants' health. Her "aery footstep" (v. 139) and "gentle feet" (v. 144), which are echoed by Maud's coming to the garden like "a dove", with "ever so airy a tread" (v. 917), stand in explicit and striking contrast to the arrival of the Red Queen in the garden of the talking flowers: "'she's coming!' cried the Larkspur. 'I hear her footstep, thump, thump, along the gravelwalk!'" (*TLLG*, p. 169).

In this sense, flowers play a powerful role of parodying and hermeneutical reversal in the Looking-Glass Land,[7] working as a manifesto of how things on the other side of the mirror follow a logic of opposition and contrast, and how human beings' importance itself is apt to relativization. This demystifying and satirical task assigned to flowers is introduced in the first *Alice* book, in the aforementioned Wonderland garden / Queen of Hearts' croquet ground. The very first sentence describing Alice's arrival there goes as follows: "a large rose-tree stood near the entrance of the garden: the roses growing on it were white, but there were three gardeners at it, busily painting them red" (*AAIW*, p. 83).[8] The purity and delicacy of the white roses is violated, in order to make them the red colour of passion and anger. This is the first sight which welcomes Alice in what she thought was a peaceful, serene garden, and turns instead to be a chaotic space of unnaturalness and homicidal wishes. The roses are therefore the first symbol of what awaits her soon afterwards. It can also be speculated that Alice herself is portrayed as a white rose turned red: after all, the pretty, innocent Victorian girl has already turned into a hungry snake, and she has showed marked predatory instincts towards the Wonderland creatures she has met (the mouse, the lizard, the birds). Here, among cards, fake flowers and absurd trials, she will finally shake off her Wonderland dream. And how? By using her *anger* against the creatures around her, changing into another furious Queen of Hearts.

Carroll's satires and inversions are, thus, not only funny. There is always a dark side to them, and the flowers prove it clearly: if both Tennyson's Maud and Shelley's ethereal garden lady die, the Rose warns Alice about her fading, and the Queen of Hearts wants to see her beheaded. The white roses are forced to signify something they do not, making visually and physically evident what the Victorians normally did with all types of flowers. Somehow then, the sense of decay and death embedded in some of the most famous Victorian celebration of gardens and flowers, appears again in the nonsense rendering of Carroll's parodic scenarios. The white roses painted red themselves can also allude to violence and blood. As Tennyson writes in *Maud*: "the soul of the rose went into my blood" (v. 882), not only hinting at a sexual meaning but also prefiguring the deathly battle to follow.

Floral Supposition and Dream-Rushes: Cognitive Processes and Emotions

Flowers perform a highly significant role in the Carrollian nonsense universes: apart from showcasing the inconsistency of traditional beliefs and literary motifs through parody and inversion, they also allow what Elaine Scarry calls "floral supposition" (2001, pp. 177–179). This means that, in the context of the reader's imagination, a key feature of the flowers is their association with the suppositional, which "explains how flowers, which seem like the subject matter of imagining rather than a part of the compositional process, in fact constitute a formal practice" (p. 177). According to Scarry, the flower "expresses the distinct quality of cognition at work in imagining" (p. 66), and the reason for this lies in a flower's peculiar characteristics which make it easy to imagine (its size, its shape, its rarity) and in its unique essence as "latency or potency" (p. 178), which mimics the nature of imagining as "only-almost-perceiving or not-quite-perceiving" (p. 179). Scarry uses as examples passages from Wordsworth and Rilke's poems, and from novels such as *Wuthering Heights* and *Madame Bovary* (pp. 158–185), in order to prove how to prompt readers to imagine motion, and persons, and other "non-floral things" (p. 173), writers very often utilize the analogy of the flower, as a powerful imagination booster. The comparison between flower and the faculty of imagination itself relies on the in-between nature of flowers, non-percipient but yet almost-percipient (p. 179), matching the essence of imagining, which is not-perceiving and yet almost-perceiving.

We can see how Carroll's flowers prove and enrich Scarry's theory about the connections between literary descriptions and flowers. Authors and poets talk about flowers, Scarry argues, because of their strong link with human imagination; in this way, making readers picturing a concept, a feeling, a situation, a movement, is easier if the desired notion is compared to a flower. When Alice enters the garden, the Rose and the Tiger Lily are trying to understand what she is, and, in order to do this, they describe her as if she was a flower, commenting on her colour, her petals, her shape. Carroll thus achieves different goals at once: on the one hand, he highlights the limits of individual perception, emphasizing that we are all restrained by our own cognitive parameters—the flowers can think only in flower-like terms. Alice becomes a flower then, and not a particularly pretty one, and the same happens with the Red Queen, who is actually a chess piece. What would be a huge ontological difference, between an object and a human being is reduced, in the flowers' perspective, to the fact that the Queen has "nine spikes" (*TTLG*, p. 169), and Alice has none. The comical dominance of the flowers' point of view over Alice's also draws attention towards the implication of non-anthropocentric ethics, as emphasized in the previous section. This feature of the *Alice* books links them to recent posthumanist frameworks: in fact, "post-humanism, as a discourse, both exposes and ironically establishes boundaries between the human and the non-human, to facilitate a dialogue as to how those very borders can become fluid" (Jaques, 2015, pp. 2–3). In this respect, Anne Kérchy argues that, in Carroll's masterpieces, "transient embodiments, shapeshiftings, elusive meanings overturn the tyrannical rule of human reason and reclaim the relevance of alternative perspectives of formerly powerless lesser life forms" (2018, p. 9).[9] On the other hand, imagining Alice as a flower, as Scarry points out, is the easiest way of imagining her: to do so allows the readers to represent in their mind the little girl who is now growing up and who is leaving the happy times of childhood behind her, and also to capture the feelings this process causes in Carroll himself. Seeing Alice as a fading flower, in a half bitter, half comical way, is the most effective way of achieving these effects.

One of the features which makes flowers so easy to visually represent in the mind is their "rarity", their airy substance which causes their association with evocation of movements (Scarry, 2001). Alice, while in the company of the White Queen, finds herself in the Sheep Shop, with the White Queen turned into the sheep. The chapter, called "Wool and Water", is all about metamorphosis, instability of substances, and mutable shapes (*TTLG*, pp. 204–217). In order to capture this, Carroll introduces other flowery plants: the "dream-rushes". Whenever Alice, who finds herself in a boat with the sheep, tries to catch some scented rushes growing near the water, she notices that there are always more beautiful ones appearing farther away, and the ones she actually manages to collect "being dream-rushes, melted away almost like snow, as they lay in heaps at her feet" (*TTLG*, p. 215). The comparison with snow adds to the elusive quality of the rushes, since snow is an unstable element, going fast from solid to liquid and then disappearing. The whole chapter is also about memory, which in the reversed world of the Looking-Glass Land works as memory of future events, and the rushes exhibit the features of beloved memories fading away like a dream.

Therefore, Carroll's dream-rushes give visible substance to the insubstantiality of memories, exploiting the transient nature of flowers (what Scarry calls "latency or potency") as a means to represent fleeting, hard-to-capture emotions, and as a way to depict the mutable, ephemeral working of memory. The flower occupies a unique position "in the arc between material and immaterial" (Scarry, 2001, p. 62), and this makes it the ideal literary device to visually elicit in the readers' mind certain transient emotions and concepts. The "almost-sentient" nature of plants and flowers can offer the possibility of describing the human mind: Carroll, once again, proves to be engaged with extremely innovative ways of conceiving the vegetative world. This aligns well with current critical plant studies (Gagliano et al., Laist) and posthumanist perspectives (Jaques), according to which the anthropocentric and zoocentric bias is overcome by directing more prominence and attention to plants, which "can no longer be portrayed as passive organisms merely subjected to environmental circumstances", but must be recognized as "information-processing organisms with complex communication strategies" (Gagliano et al., 2017, p. xiv).

Dark Woods as Lost Paradises

The mysterious space of woods and forests has always represented a meaningful trope in children's literature. Jack Zipes, analysing in particular the wood in the fairy tale of Little Red Riding Hood, appropriately states "woods are the natural setting for the fulfilment of desire. The conventions of society are no longer present. The self can explore its possibilities and undergo symbolic exchanges with nature inside and outside the self" (1993, p. 361). If several traditional fairy tales stage the presence of woods as a symbol for the unfamiliar, and as a setting of transformation, as well as a site of potential danger (such as Little Red Riding Hood, Hansel and Gretel, The Snow Queen, Snow White, The Sleeping Beauty), in Victorian children's stories it is in particular the wildness of woods, as opposed to the domesticated space of the garden, that represents an ever-recurring theme, linked in particular to the possible peril hidden in it.

In this sense, Christina Rossetti's *Goblin Market* (1862) and *Speaking Likenesses* (1874) both introduce the forest as a place where temptation, sin and monstrosity lurk in dark recesses; similarly, in *The New Mother* by Lucy Lane Clifford (1882), the woods surrounding the cottage where the characters of the story live represent disobedience and untamed behaviour. George MacDonald's forests (for examples in *Phantastes*, 1858, *The Golden Key*, 1867, *At the Back of the North Wind*, 1871) are more complex and heterogeneous spaces, always appearing in his fairy tales, and marked by the presence of magic, both evil and good.[10] As with gardens and flowers, then, once again Carroll had a rich background with which to work and his *Alices* show a rather unique take on the role of woods in children's literature.

After having run away from the White Rabbit's house (in a comical reversal of her initial running *after* him), Alice finds herself "safe in a thick wood" (*AAIW*, p. 45). Within this initial description, another of Carroll's parodic inversions is apparent: the danger for Alice is in the pretty little house of the White Rabbit, and safety lies in the thick dark wood. What is particularly interesting in Carroll's worlds, considering the general Victorian perspective on woods in fairy tales, is that, while gardens are shown to be places of strangeness and folly, woods are depicted as safe areas of peacefulness. Alice encounters a "dear little puppy" (*AAIW*, p. 47) in the Wonderland wood who, despite being big in comparison to Alice's diminished size, is the only "normal" creature Alice meets in Wonderland, the only one who behaves exactly as a puppy might behave in the real world. Afterwards, she meets the Blue Caterpillar standing on the top of a mushroom who, even if he is another of Wonderland's queer creatures, also works as an exception: if, in Wonderland, there is a constant lack of communication and understanding between Alice and the place's inhabitants,[11] the Caterpillar can understand Alice without even the need for her to speak. He can, in fact, read her mind (*AAIW*, p. 55).[12]

Wonderland wood, then, paradoxically functions as the only safe place in Wonderland's chaotic madness, where communication is possible and where some traces of the "normal" world can still be found. The dark forest described in *Through the Looking-Glass* plays a similar role: the two *Alice* books are obsessively inhabited by language games and puns, and the only moment of quietness, silence and harmony happens in the "woods where things have no name" (*TTLG*, pp. 185–187). Here Alice cannot recall her own name, nor the name of anything around her, and this enables her to experience "a quasi-Wordsworthian communion between the human and the natural worlds which is unlike anything else in the *Alice* books" (Haughton, 2003, p. 338). A new way of experiencing the vegetative world, Carroll seems to suggest, can exist only outside the frantic taxonomic obsession of the Victorians, where names and meanings were arbitrarily assigned to plants in order to establish a supposed human superiority.

In this happy, nameless place, Alice meets a Fawn, and, because they are both blissfully ignorant of their own identities and separate categories, they embrace and walk thus through the forest. As Beer underlines, here "a momentary paradise prevails, the moment of paradise *before* Adam named the animals and kinds" (2016, p. 166). Again, a posthumanistic perspective is here celebrated, in the peaceful woods where the violence of human linguistic classification has not intruded and where it is instead proposed "an alternative, non-discursive, nonsensical classification based on empathic interspecies relationality" (Kérchy, 2018, p. 16). Zipes's "symbolic exchanges with nature" are here taken to their extreme possibilities, towards a proper symbiosis between human and natural dimensions. Hence, Carroll reworks the fairy tale trope of the enchanted woods, giving it a stronger sense of detachment from social conventions (the most basic ones of language) and placing it in opposition to the more popular Victorian representation of forests in children's stories as locations embedded with negative undertones of immorality and danger.

Conclusion

Gardens, flowers and woods as represented in Carroll's *Alice* books have been analysed in turn and disclosed as narrative sites of important meaning: Carroll utilizes vegetation in his fictional worlds in order to discuss and question Victorian literature's traditions and stereotypes, and in order to promote new semantic possibilities, discussing anthropocentric assumptions. Normally considered as a domesticated space, the Victorian garden can actually conceal death, madness and irrationality among its beautiful flowerbeds; the language of flowers can convey quite different meanings to the ones expected; and the dark woods can turn into Edenic places of harmonic balance. In this sense, Carroll's *Alices* provide a new perspective on the multifaceted possibilities vegetation (in the form of gardens, flowers, or forests) can offer for Victorian (and not only Victorian) children's literature.

Focusing on the role of vegetation in the *Alice* books can highlight in new ways Carroll's strong critique of Victorian society, and can function as a means of promoting plant-based, posthumanist viewpoints in which anthropocentric biases are questioned and "radical destabilizations of hierarchies of beings" (Jaques, 2015, p. 5) are put in place.

Notes

1 Research for this chapter was supported by the European Regional Development Fund (Mobilitas Pluss, MOBJD491) and the Estonian Research Council (Grant 1481).

2 Gillian Beer, for instance, highlights that knowledge of botany has informed Carroll's take on the depiction of the talking flowers (2016, 150–153), while many critics have remarked upon the mockery of Victorian ideas on the language of flowers Carroll perpetrates in the same chapter, especially with reference to the parody of Tennyson's *Maud* (see Carroll/Gardner, 2001; Shires, 1988; Rackin, 1997).

3 Illustrators also contributed to the development of this idea of little girls in delightful flowery gardens in the Victorian imagination: for instance, Kate Greenaway in her *Under the Window* (1879) portrayed many scenes of garden tea parties, in which girls with "curly golden hair, big blue eyes, merry smiles and rosy cheeks" (Carpenter, 1985, p. 107) take their tea in the setting of pleasant Victorian gardens.

4 It is also interesting to note how in both quotations there is the recurrence of the word "garden" next to the verbs "to wander" and "to wonder", which are key verbs behind the creation of Alice's adventures: Wonderland is the place where one wanders in childhood's dreams, in a constant state of wonder. The importance of the role of the garden is thus reinforced by being placed in connection with these two revealing words.

5 The fact that in both Alice's stories we are in a dream-world does not undermine the importance of the messages here conveyed: on the contrary, Carroll assigned an extremely relevant role to the dream-dimension. He was an active member of the Society for Psychical Research, for which dreams were carriers of profound, deep, and revelatory meanings. Moreover, what Carroll continuously stresses in his *Alice* books is how the boundaries between reality and dreams are difficult to be rigidly drawn: in this sense, the nonsense of Alice's dreams reflect the nonsense of Alice's reality—and the other way around.

6 Carroll's innovative perspective on how little girls should or should not behave is shown also in the way he preferred Alice to be depicted: as Kiera Vaclavik has emphasized in her *Fashioning Alice*, he did not want Alice to be entrapped in uncomfortable sophisticated garments. Instead, Alice's outfit was meant to be a very common one, one which could have allowed her some freedom of movements, the chance to play around (67–70). Moreover, looking at the history of Carroll's photographic endeavours, he always wanted the girls he portrayed to have fun, to wear comfortable clothes, in order to be able to enjoy themselves in the countryside, not to be just clean and elegant little ladies in a closed garden.

7 It is true that the main target of Carrollian irony was most likely the Victorian age and its traditions and most-celebrated literary works, but it is also important to recognise how "talking flowers [...] have appeared for centuries in folk literature and predates Carroll's use of them in children's literature by decades" (Reichertz, 2000, p. 14)—which shows how Carroll was parodying literary symbols and tropes of folktales and traditional stories for children.

8 Another possible speculation about the meaning of this episode links it to a parody of the War of the Roses, with the Red Rose symbolising Lancashire and the White Rose representing Yorkshire.

9 For more on Victorian writers and posthumanism, see in particular two special issues of CVE: "Comforting Creatures", and "Becoming Animal".
10 For more on George MacDonald's use of woods and forests, see "Narrative Strategy: The Role of Woods and Forests in George MacDonald's Fairy Tales", by Suneeti Chhettri Lock.
11 An aspect of the Carrollian worlds widely recognized by critics; see for instance Douglas-Fairhurst, 2015.
12 About the Caterpillar's mind-reading skills and their possible connections to Carroll's affiliation with the Society for Psychical Research, see Gardner, 2001, pp. 55–56.

References

Arniac, A., Cingal, D., Courtney, J., Crampe, F.*et al.* (Eds.). (2018). Comforting creatures [Special Issue]. *Cahiers victoriens et édouardiens*, 88.

Beer, G. (2016). *Alice in space*. Chicago, IL: The University of Chicago Press.

Bilston, S. (2008). Queens of the garden: Victorian women gardeners and the rise of the gardening advice text. *Victorian Literature and Culture*, 36(1), 1–19.

Burnett, F. H. (2011). *The secret garden*. London: Penguin Books. (Originally published in 1911.)

Carpenter, H. (1985). *Secret gardens: A study of the golden age of children's literature*. Boston, MA: Houghton Mifflin.

Carroll, L. (2001). *Alice's adventures in wonderland*. In M. Gardner (Ed.), *The annotated Alice* (pp. 7–132). London: Penguin Books. (Originally published in 1865.)

Carroll, L. (2001). *Through the looking-glass and what Alice found there*. In M. Gardner (Ed.), *The annotated Alice* (pp. 137–288). London: Penguin Books. (Originally published in 1871.)

Cohen, M. N. (1989). *The selected letters of Lewis Carroll*. London: Palgrave Macmillan.

Douglas-Fairhurst, R. (2015). *The story of Alice: Lewis Carroll and the secret history of wonderland*. London: Harvill Secker.

Engelhardt, M. (2013). The language of flowers in the Victorian knowledge age. *Victoriographies*, 3(2), 136–160.

Gagliano, M.*et al.* (2017). Introduction. In *The language of plants: Science, philosophy, literature* (pp. vii–xxxiv). Minneapolis: University of Minnesota Press.

Green, R. L. (Ed.). (1971). *Lewis Carroll's Alice's adventures in wonderland and Through the looking-glass*. Oxford: Oxford University Press.

Haughton, H. (Ed.). (2003). *Lewis Carroll's Alice's Adventures in Wonderland and Through the Looking-Glass*. London: Penguin Classics.

Jaques, Z. (2015). *Children's literature and the posthuman: Animal, environment, cyborg*. New York: Routledge.

Kérchy, A. (2018). Alice's non-Anthropocentric ethics: Lewis Carroll as a defender of animal rights. *Cahiers victoriens et édouardiens*, 88, 1–35.

Laist, R. (Ed.) (2013). *Plants and literature: Essays in critical plant studies*. Amsterdam and New York: Rodopi.

Lock, S. C. (2010). Narrative strategy: The role of woods and forests in George MacDonald's fairy tales. *American, British and Canadian Studies*, 15, 148–161.

MacDonald, G. (2008). *Phantastes*. London: Pater Noster. (Originally published in 1858.)

Moine, F. (Ed.). (2017). Becoming animal [Special Issue]. *Cahiers victoriens et édouardiens*, 85.

Rackin, D. (1997). Mind over matter: Sexuality and where the "body happens to be" in the *Alice* books. In Lefkovitz (Ed.), *Textual bodies: Changing boundaries and literary representations* (pp. 161–184). New York: State University of New York Press.

Reichertz, R. (2000). *The making of the Alice books: Lewis Carroll's uses of earlier children's literature*. Montreal : McGill-Queen's University Press.

Rossetti, C. (2013). *Goblin market*. Alcester: Pook Press. (Originally published in 1862.)

Scarry, E. (2001). *Dreaming by the book*. Princeton, NJ: Princeton University Press.

Seaton, B. (1985). Considering the lilies: Ruskin's 'Proserpina' and other Victorian flower books. *Victorian Studies*, 28(2), 255–282.

Seaton, B. (1995). *The language of flowers: A history*. Charlottesville, VA: University of Virginia Press.

Shires, L. M. (1988). Fantasy, nonsense, parody, and the status of the real: The example of Lewis Carroll. *Victorian Poetry*, 26(3), 267–283.

Tennyson, A. L. (2007). *Selected poems*. London: Penguin Books.

Vaclavik, K. (2019). *Fashioning Alice: The career of Lewis Carroll's icon, 1860–1901*. London: Bloomsbury.

Yoshida, A. (2013). Temptation of fruit: The symbolism of fruit in Christina Rossetti's "Goblin market" and in the works of D. G. Rossetti and J. E. Millais. In R. Laist (Ed.), *Plants and literature: Essays in critical plant studies* (pp. 75–84). Amsterdam and New York: Rodopi.

Zipes, J. (Ed.). (1993). *The trials and tribulations of little red riding hood*. New York: Routledge.

5 Fern Blossom and Lilibala

Magical Plants in Serbian Children's Fantasy Fiction

Tijana Tropin and Ivana Mijić Nemet

In this chapter we provide an overview and an analysis of fantastic and magical plants (that is, of their heterogeneous origins and functions) in contemporary Serbian fantasy fiction written for children, and combine a traditional analysis of the literary fantastic and folklore elements with eco-criticism. Serbian folklore beliefs concerning magical plants and their properties were adopted and made use of by various children's authors. Older authors rely heavily on the aforementioned folk heritage, using magical or fantastic plants as signals of the supernatural order and symbols of higher ethical or religious values. Recent authors, on the other hand, show a marked tendency to employ plants of their own invention in order to convey an ecological message. We focus on the following texts, which demonstrate this development of the plant motif in Serbian children's fantasy: Desanka Maksimović's literary fairy tale *Patuljkova tajna* [The Dwarf's Secret] (1932), Grozdana Olujić's literary fairy tale *Mesečev cvet* [Moon Flower] (1979), Uroš Petrović's novel *Aven i jazopas u zemlji Vauka* [Aven and his Badger-Dog in the Land of Vauks] (2003), Mina Todorović's trilogy *Virovi* [Vortices] (2003–2016), Ivana Nešić's novel *Zelenbabini darovi* [Green-mother's Gifts] (2013), and Selina Lovren's short novel *Malo zeleno drvo* [The Little Green Tree] (2019). The first part of our examination portrays the shift in the selected narratives from an organic (i.e. folklore-based) relationship between protagonists and plants to an invented one (presenting a belated response to the environmental crisis). The second part focuses on the fantasy of the Anthropocene vs. fantasy in the Anthropocene, specifically on the negotiation between an anthropocentric and an ecocentric horizon. Finally, the third part examines recent trends and their response to the environmental crisis, as well as a recursive shift towards folklore motifs that involves a gently parodic turn.

We have highlighted how the different trends developed over time by contrasting the selected works in pairs: Maksimović's fairy tale vs. Olujić's, Petrović's novel vs. Todorović's, and Nešić's novel vs. Lovren's. We also adopt a hermeneutic approach and provide a close reading of the selected corpus, with special emphasis on the issue of the fantastic and magical plants, deconstruction of their anthropomorphic perception, and change in

DOI: 10.4324/9781032066356-8

the way plants are represented in children's fantasy texts: from the Aristotelian notion of a species hierarchy to the most recent notion of "plant intelligence" (see Guanio-Uluru, 2018, p. 127).

Regarding the theoretical framework, we turn primarily to recent ecocritical and posthumanist theory, especially as espoused by authors who use this viewpoint to approach children's fantasy literature (such as Bradford et al., 2008; Jaques, 2015; Attebery, 2018; Oziewicz, 2018a; Oziewicz, 2018b; Goga, 2018; Chen, 2018). These researchers destabilize the anthropocentric humanist ideals via posthuman theory and challenge commonly accepted sovereignty of humanity over nature. They have devoted increasing attention to the plant kingdom and included the vegetal in their study of children's and YA fantasy. Their insights offer the possibilities of a new approach to Serbian children's fantasy, which is usually analysed with a focus on contemporary Western literary influences and not within a theoretically based ecological framework (as we shall see, the single exception being the texts of Uroš Petrović). Ecocritical theory offers the potential of a fresh reading which highlights some neglected characteristics of the selected texts.

The presence of plants in songs, tales, and legends in areas where Serbs live today can be traced back to the first records of Serbian oral literature. Vuk Stefanović Karadžić's *Rječnik* [Dictionary], an unorthodox encyclopaedia of Serbian folk life, beliefs, culture and history, contains over nine hundred lexemes that refer to real and imaginary plants, and their names are largely motivated by the human image of the world. Humanity's (anthropocentric) view of its most immediate environment has a strong impact on the denominations of flora, so plants are usually named according to colour, scent, taste, habitat, parts of day, weather, their healing or magic properties etc. Personification is also very common in floral names, indicating strong links between humans and plants (compare Karanović & Dražić, 2016). Serbian oral poetry and oral literary forms in general also demonstrate anthropocentrism in their manner of plant depiction. Plants and herbs are mainly present in the ritual-mythical and magical layers of oral lyrical poetry (compare Vukmanović, 2016), while fairy tales single out important plants only, and rarely mention them if they do not carry a particular meaning or function (see Radulović, 2009). Both poetry and prose mainly refrain from describing plants, merely naming them instead, in keeping with the genre norms of verbal folklore. They are usually depicted as wonderful gifts with magical features or magical powers used to overcome spatial obstacles. Accordingly, their main function is the advancement of the plot.

From Romanticism to the present, oral literature has been a source of fruitful inspiration for Serbian written literature. While floral motifs transposed from traditional oral culture are a strong presence in recent and contemporary Serbian literature, they are interpreted and used differently. When creating imaginary flora, contemporary writers of fantasy for children and adults rely on both oral literature and their own imagination, and often their writing has a decidedly ecological slant.

Modern Serbian children's literature started developing relatively late, in the second half of the 19th century, and leaned heavily on two sources: on the one hand, the contemporaneous European (usually German) children's literature, including translations, adaptations and imitations of genre models, and, on the other, the Serbian folk heritage. This means that the motif of magical plants appears relatively early: the two best-known examples are *raskovnik* and the fern flower. Raskovnik (lit. "chain-breaker"), an imaginary herb that has the power to break chains, open locks and, less commonly, bring prosperity and money to its finder (Tolstoj & Radenković, 2001), has often been used as an important symbol in poetry and literature for adults (e.g. Borislav Stanković's short story *Biljarica*, which describes a woman's lifelong search for the magic herb; see Jokić & Marićević Balać, 2018). The fern flower, or alternatively the fern seed, is another version of the magical herb. Allegedly, it enables its finder to see treasure hidden beneath the ground, comprehend the language of animals, gain universal knowledge (including such trivia as what the emperor had for dinner the previous night) or become invisible. The catch in this legend is, of course, the fact that ferns, being vascular plants, have neither flowers nor seeds. The majority of fern-related folk beliefs claim that they only blossom once a year, on Saint John's Eve (Tolstoj & Radenković, 2001, pp. 416–417), and that on this night they are often guarded by forest spirits such as vilas, or by the devil himself.

Fern Blossom and Moon Flower: The Magical Flowers in Fairy Tales

As we shall see in this section, the early use of fantastic plant motifs in children's fiction is inextricably linked to the fairy tale, though this genre later moved away from its folkloric roots. An important example of the fern flower's appearance in fiction is Desanka Maksimović's short story *Patuljkova tajna* [The Dwarf's Secret], first published in 1932. Maksimović (1898–1993) is probably the most important Serbian 20th-century female poet and children's author, and her opus spans seven decades. This story was included in her most important post–war story collections, *Ako je verovati mojoj baki* [If my grandma is to be believed] and *Patuljkova tajna* [The Dwarf's Secret]; originally written and published before World War II, it presented a significant contrast with realist fiction, which marked the 1950s in Serbian children's literature.[1] It centres on the figure of an unnamed old dwarf, who spends a day walking through the forest encountering various beings, such as birds (a wood grouse), trees (an old pine tree) and animals (a viper). In each of these encounters, he notices and recognizes a sign that the creature in question will die on that same day, which compels him to consider his own mortality. He attempts to rejuvenate himself by drinking the "living water" from a forest spring (another well-known folklore motif). Failing this, and having recognized a portent of his own impending death, he finds a little girl and tells her his most treasured secret: how to find the fern blossom and be happy for the following year, or possibly forever.

Maksimović subtly shifts the balance in this story. Unlike the typical oral fairy tale, this one is no longer concerned primarily with the hero's attempts to gain the magic flower and make use of its powers—on the contrary, as the readers follow the dwarf on his leisurely walk, they become aware both of his overwhelming fear of death as well as of his secrecy, and his habitual, selfish concealing of his accumulated knowledge. The dwarf's final action, however, is a last desperate attempt to convey at least one important piece of information. The little girl, conversely, shares this secret—and thus the happiness that the golden fern flower brings—with others.

It is clear that the author is not concerned with the fantasy element as such, seeing as how within the narrow scope of this story neither the flower's appearance, nor its powers are described in detail. The flower functions as a symbol of fleeting happiness and, more importantly, a catalyst for the characters' actions, highlighting their personalities. It fails to bring any solace to the dwarf, but the generous little girl and her shepherd friends will be happy forever. In the background of this fairy tale, however, another story takes shape. The wood grouse is killed by hunters, while the snake is killed by those very shepherds. The relationship between the forest and humans, as Maksimović shows us, is far from idyllic.

An analogous development of the magic flower motif can be observed in a more recent fairy tale: Grozdana Olujić's *Mesečev cvet* [Moon Flower]. Olujić's unusual and lyrical fairy tales combined a heightened awareness of modern-day childhood in an urban environment with highly stylized prose and an exceedingly wide-ranging use of literary motifs, including a strong link to Indian heritage and Buddhism (see Opačić, 2011). *Mesečev cvet* is representative of the more recent tendencies in Yugoslav and Serbian children's literature. Although it focuses on an old motif, the search for a magically healing flower, the boy Vedran is not a typical fairy tale protagonist. As a sickly city child, raised in "glass towers", Vedran neither knows nor respects nature; during his search, he tears down a bramble bush, destroys an anthill, and finally pollutes a wellspring. Such behaviour is not in accordance with the Romantic idea of the child being uniquely related to nature, and reflects modern alienation. Vedran is duly punished by the mythical figure of the Master of Herbs, and he has to atone for his sins by serving the forest plants for three years. His sentence does not pass in vain: Vedran learns the language of the plants, grows closer to the living beings of the forest, and even strikes up an unlikely friendship with them. The chestnut tree explains where and how to find the Moon Flower, which, being even rarer than the fern flower, blossoms once in a century. The bramble bush, the ants, and the water all help him reach his goal. Vedran, however, passes the ultimate test by *refusing* to pick the beautiful flower. His time of servitude has taught him to respect and love nature, and to cooperate with plants and insects. His newfound health, Olujić signals strongly, has been caused by the overall change in his behaviour and his lifestyle. Vedran has to learn from all living beings in order to grow and develop into a responsible adult. The ecological subtext of Maksimović's fairy tale becomes an explicit message: Vedran's alienation and remoteness from nature are pinpointed as the reasons behind his illness and erstwhile unkindness towards others.

Olujić's image of the Moon Flower is much more detailed than Maksimović's fern flower: "Delicate and silver, the Moon Flower grows into the sky like a giant lily, its stalk shimmering and trembling ... Delicate and lonely, it shines like a star" (Olujić, 2011, p. 42, translation ours). Olujić introduces a crucial twist to the enchanted flower motif—rather than a simple magic object, the Moon Flower is a sentient being, and Vedran instinctively recognizes that picking the flower is tantamount to killing a human. This is further underlined by the fact that the flower addresses Vedran as "little brother" before parting, reinforcing the sense of kinship between all living things. Arguably, Olujić's literary fairy tale functions "as an imaginative platform for running speculative thought experiments about being human in a world that is becoming increasingly out of sync with its own ecological and biological rhythm" (Chen, 2018, p. 181). Its shift away from the folkloric roots so prominent in Maksimović's tale is also important as a herald of the following generation of fantasy writers.

The Advent of Fantasy Proper: *Aven* ... and the *Vortices* Trilogy

In this section, we shall see how the writers of the first Serbian children's fantasy novels dealt with ecological themes and plant motifs in the early 2000s. The previously described ecological awareness became increasingly more frequent in Serbian children's literature as time progressed. This reflects various factors: first, the rise of environmental awareness that gained global momentum in the 1970s and 1980s slowly entered Serbian literature, which meant industrialization was no longer seen only as a reflection of progress but also as a condition with its own problems (such as pollution and the devastation of suburban landscapes); second, the turbulent 1990s marked a breach with the previous literary trends and their gradual evolution. Many children's authors retreated into writing fantasy or historical narratives that bore little or no overt relation to the current reality of war, poverty and social upheaval. Ecological motives were an easy choice for realist narratives, as the authors could simultaneously claim a relevant topic and avoid much more problematic themes. During the same period, some fantasy writers included mention of fantastic or miraculous herbs and trees (Rosić, 1991; Milivojević, 1997; Velmar-Janković, 1998), almost without exception framing them within the fairy tale genre, often in a medieval context.

The following period of the early 2000s was marked by an intense interest in fantasy for children, and a new generation of writers, influenced by diverse Western fantasy authors, appeared on the Serbian literary scene.[2] Probably the most popular and influential writer to emerge in this period is Uroš Petrović. His first novel, *Aven i jazopas u zemlji Vauka* (2018) [Aven and his Badger-Dog in the Land of Vauks], is a meticulously developed imaginary world, in line with Wolf's statement that:

> The most common type of invention regarding an imaginary world's natural realm is that of new flora and fauna ... When invented flora and fauna are more than merely windowdressing or replacements for Primary World animals that serve a similar function (such as pets or beasts of burden), they are usually used to solve world–building problems ... Occasionally an invented plant or animal even provides the impetus for a story, as any fantasy quest to defeat a dragon demonstrates. Worlds that are subcreated to an even deeper level include new kinds of biology, ecosystems, and planets with unusual material compositions.
>
> (Wolf, 2012, p. 172)

Indeed, Petrović (similarly to Mina Todorović) creates a plethora of fantastic plants and beasts, foregrounding them systematically: most of the novel's chapters start with encyclopaedically structured descriptions of various biomes, including illustrations, binomial names for each listed species and an account of their most important characteristics. Petrović sometimes mentions extant living beings and contrasts them with their fictional counterparts; at other times, he creates imaginary flora and fauna that bear a strong resemblance to certain real-world species, but with important differences. His illustrations hark back not to the luscious illustrations of fairy tales, but to the more precise drawings in children's textbooks, presenting the fictitious images with a meticulous exactness.

The Baraba tree is the most important example of imaginary flora presented in the novel, as well as its central symbol. It is introduced as a tree that sheds its sharp, rust-coloured leaves every day, thereby killing all living creatures in its secluded, windy valley. Here, one can draw parallels with monstrous plants within fantasy and horror fiction (e.g. Old Man Willow in J. R. R. Tolkien's *The Lord of the Rings* or the Wood in Naomi Novik's *Uprooted*). Dawn Keetley locates the horror of plants both in their absolute strangeness and in their uncanny likeness to humans: "At its most basic, plant horror marks humans' dread of the 'wildness' of vegetal nature - its untameability, its pointless excess, its uncontrollable growth" (Keetley, 2016, p. 1). The Baraba embodies an absolute alterity and refuses to be mere backdrop while subverting the "taken-for-granted 'fact' of plants' invisibility, passivity, and harmlessness" (Keetley, 2016, p. 10).

On the other hand, although described as an ominous presence from the very start, it gradually becomes clear that this plant is actually beneficial for humans, as it keeps them safe from the threat of Vauks. Here malice is an outside influence, and not in the tree itself. As Aven finds out, the tree was originally a chestnut, transformed into the Baraba by a magic spear hurled into its roots (Petrović, 2018, p. 273). After Aven removes the spear, the tree reverts to its original, benign form. In a symmetrical act, performed during the last desperate battle against the Vauks, Aven turns a "large aucuba tree" into another Baraba, which promptly starts shedding its leaves.

The origin and function of the Baraba tree are representative of Petrović's attitude towards nature in general. While conceding that humans are part of the world's ecosystem, Petrović focuses his attention on the various practical *uses* of his flora and fauna. More often than not, his descriptions of the various species position them as useful or detrimental for humans. As Keetley succinctly put it: "Recent philosophers of plant life have routinely noted humans' long–standing instrumental attitude toward plants. Indeed, when our blindness wavers, when we actually see plants, we see only something made to serve us" (Keetley, 2016, pp. 19–20).

In her discussion of fictional plants, Jaques notes that the relationship between human and tree is often coached in terms of mastery (Jaques, 2015, p. 117). *Aven* provides a clear confirmation of this thesis: from the start, Aven is singled out from his peers because he is the only boy who can *use* Baraba's leaves, throwing them with a deadly precision. His attitude towards the tree is not marked by respect or tenderness; the tree is an adversary to be overcome. Even the healing of the tree is not driven by selfless impulses—Aven acts in accordance with a prophecy, and in order to help his side win the war. This is a story about mastering nature. Everything is summoned to serve the human being (compare Røskeland, 2018). According to this, Petrović's novel can be classified as a "fantasy of the Anthropocene", naturalizing human dominance, exploitation, and domestication of the non-human world (see Oziewicz, 2018b).

Serbian authors usually classify *Aven* as an early and powerful representation of *ecological fantasy*, offering convincing arguments in support of that thesis (see Pešikan Ljuštanović, 2009; Jocić, 2014; Glavinić, 2016 throughout). Still, Petrović's anthropocentrism is a noticeable and important part of the novel, and any ecocritical reading of *Aven* has to consider the suggestion to read "against the grain of foregrounded interests" (Mallan, Stephens, & McCallum, 2008, p. 84). Since it is based on firm Anthropocene foundations, Petrović's fantasy doesn't change the way we see reality, nor does it offer a reimagining of our impact on nature (see Attebery 2018; Oziewicz 2018b).

As Glavinić notes, it is easy to categorize his protagonists with the help of binary oppositions such as healthy/ill, good/evil, natural/unnatural[3] etc. Such a simplified view can partly be explained by Tolkien's strong influence on Petrović, who models his antagonists, the Vauks, in large degree after Tolkien's Orcs, displaying a similar disregard for psychological complexity. An important exception to the simplistic binary divisions is the Baraba tree,[4] which unites unadulterated nature (as represented by the perfection of the gingko tree, see Glavinić, 2016), i.e. the good and the natural side of the equation on the one hand, with illness and deformation on the other. In the same vein as Tolkien's Old Man Willow and Novik's Wood, Baraba allows its sorrow and pain to turn into malice, hate, and destruction. But while

Tolkien's and Novik's plants lash back at the human destroyer, Petrović's narrative depicts the effects of Baraba's corruption as something beneficial for the human race and not as a revenge of the oppressed.

Mina Todorović, whose first fantasy novel *Vir svetova* [The Vortex of the Worlds] appeared in 2000, offers a different approach. Akin to Petrović, she created a world inspired primarily by the Western fantasy tradition, especially Tolkien and, to a certain extent, C. S. Lewis (see Tropin, 2012). Todorović's complex world-building, however, displays an important shift away from anthropocentrism, and ecological awareness is more clearly present (albeit introduced more clumsily) compared to Petrović's works. Unlike her contemporary, Todorović squarely places the blame for the ecological crisis on humankind and its greedy exploitation of natural resources. Her fantasy worlds, especially the one introduced in the first part of the trilogy, represent a clear contrast to our own, and the characters stress this time and again. This is in keeping with Oziewicz's views on the potential of children's fantasy literature to challenge the fantasy of the Anthropocene: "The battle for our planet's future is first and foremost a battle of imagination, in which stories are the most potent means to rewire our cognitive and affective modes of being to resist ecocide and engineer a positive change" (Oziewicz, 2018b).

From the very beginning of the first novel, Todorović establishes a firm link between her human characters and nature, especially trees. An enormous oak tree turns out to be a portal or "vortex" between the worlds, existing both in the primary and the secondary world. It represents the healthy, integral nature such as has become nearly non-existent on Earth. Two elderly botanists are by chance transposed into the secondary world and, almost immediately, sent on a quest. Todorović's worldbuilding is, as mentioned, influenced by Tolkien, but it shows vestiges of Slavic mythology as well (see Tropin, 2015). Her imaginary flora and fauna, on the other hand, show a distinct originality.

Todorović's most important representative of magical flora is the *lilibala*, a legume-like plant which bears bitter seeds that resemble chick peas (often called *leblebija* in Serbian). Lilibala produces a small number of seeds once every five centuries, and these seeds are coveted by all races of the secondary world: humans, elves, the long–nosed Pis people and the wizards alike. As one character explains, "lilibala gives to everyone what they need most …" be it longevity, wisdom, or even immortality. Todorović's botanists confuse two lilibala seeds with ordinary chick peas and consume them, instantly gaining the understanding of all languages and superhuman longevity. Their quest is then motivated by a desire to help two of the Pis people, who are doomed to death if a replacement for the lilibala is not found. The lilibala is thus primarily a multifunctional plot device, denoting the infinite possibilities of the magical secondary world. Still, Todorović draws a parallel with other rare and endangered plant species, stressing that lilibala is rare because of the great demand for it. As we find out in later volumes, the lilibala has a dark counterpart, the poisonous *gurbigaba*. Gurbigaba is presented as a typical invasive species, overpowering the lilibala and replacing it with its own seedlings (Todorović, 2004, p. 126). Todorović finally links both plants to the same source, the mighty female dragon, appropriately named *Magma*. She lives in the heart of the world and her wrath or joy are linked with the state the world is in and, more particularly, with the existence of lilibala and gurbigaba. In a reworking of end-time myths, Magma is expected to destroy the world if she becomes wrathful enough—namely if all the existing lilibala plants die. The final quest of the botanists ends (of course) with the victory for the Good and the triumphant appearance of new lilibala seeds, a gift from the dragon. The two plants are shown not to be mere polar opposites, but inextricably connected as well, and their alternative names—life-giver and death-giver—point to Todorović's all-encompassing vision of life and death, surpassing the obvious ecological messages.

A third world appears in the third book, created by the chief antagonist, the evil wizard Crmas. Named "the desperate world", it is devoted entirely to the suffering of its human-like inhabitants, who spend their short lives by toiling in factories surrounded by a barren wasteland. The only "tree" in that world is a dead plastic plant, decorated with ribbons instead of leaves. This darkly satirical image is depicted with an urgency and intensity which are markedly distant from Todorović's early work, reflecting the dramatic changes in the environment in recent decades. Her vision of an entirely artificial, industrial world bereft of nature incorporates radicalized, grotesque images of acid rain and deforestation on one hand, and childlike factory workers that are forbidden to play or tell stories on the other. This drastic reimagining of our own world is a call for action: within the *Vortices* trilogy, its untenable position is resolved by a combination of magic and personal sacrifice, as plants and animals are introduced to the "desperate world", creating a promise of a balanced, happier environment for its inhabitants. However, Todorović's awareness of human culpability remains far more pronounced (and more nuanced) than Petrović's.

Raskovnik and the Little Green Tree: Ivana Nešić and Selina Lovren

Our final contrastive analysis deals with a pair of recent children's novels that use folklore heritage in a far more playful spirit than their predecessors. The youngest author included in this analysis is Ivana Nešić. Her first novel, *Zelenbabini darovi* [Greenmother's gifts] (2013), is characteristic of her approach to traditional tropes and folklore motifs. Irreverent, zany humour abounds, but its subversive potential is concealed within a seemingly standard quest narrative. Mika, a lonely little boy, attempts to help a lost, masterless tribe of household dwarfs (*maljutak*), and is tricked into performing services for the eponymous Zelenbaba, apparently a powerful witch. Nešić's novel starts out with a realistic depiction of Yugoslav provincial life in the 1950s, but then the fantastic element intrudes and Mika realizes that his world is populated with beings from Serbian and Slavic folklore: household dwarfs, talking animals, witches, as well as more sinister figures such as *čuma* (the personification of the plague), or the murderous water-sprite, *vodenjak*. This impressive repertoire includes two important mythological plants: the herb *raskovnik* and the world tree. Nešić uses both motifs simultaneously in order to advance the plot (Mika has to pluck a fruit from the world tree and, later, to find a *raskovnik* plant and open a magic passage), subjecting them to gentle parody in the process as well. Mika is apt to interrogate the mythical images from a rational point of view; for instance, he wonders how the world tree can hold the entire world together and not be visible to Earth's inhabitants, especially since the Earth is round anyway. Unfortunately, he discusses the question while climbing the tree itself. Correspondingly, his attempts to procure a *raskovnik* "by the book", i.e. by obeying the traditional rules and trapping a hedgehog or turtle younglings, are met with derision from the animals, who simply refuse to follow the outmoded stories. However, the magic herb itself is described without a trace of irony (Nešić, 2013, p. 141); not only iridescent and unique, it also inspires both animals and humans with a wish to protect it from evil and to prevent its misapplication. While *Zelenbabini darovi* does not include any overt ecological message, Nešić discreetly develops the themes of responsibility, especially for one's environment, and the interdependence of all living beings. Perhaps the most telling botanical detail is the contrast between an old and always welcoming pear tree in Mika's neighbourhood, described in the beginning, and the sickly, underdeveloped city trees in a later section of the novel. Nešić's novel subtly seeks to cultivate an appreciation for nature in the young readers and calls for humankind to find a balance with the natural world.

Unlike Nešić, Selina Lovren has published only one children's book to date, *Malo zeleno drvo* [The Little Green Tree] (2019). Although this short novel has yet to receive critical attention, it is a good example of the evolution of the magical plant motif and its gradual linkage to the rise of ecocriticism and environmental awareness. Lovren's protagonist Maja wishes to cure her mother, who is suffering from a mysterious illness. As it turns out, her mother is inexplicably linked to an old oak tree at the centre of the nearby forest, and all three, the woman, the tree and the forest, are equally afflicted. Maja embarks on a rather stereotypical quest, gathering several magic objects, gaining friends and allies, before succeeding in saving the oak and consequently her mother.

According to Le Guin's *The Carrier Bag Theory of Fiction* (1986), Maja can be described as a hero collector, one who collects transformative experiences through her encounters. Thus, Maja matures and gains the ability to comprehend the whole picture and to save her mother and the forest (compare Nina Goga's analysis of Frida Nilsson's ecofantasy *Ishavspirater* in Goga, 2018).

The most original and powerful symbols in Lovren's narrative are the "little green tree" which follows Maja everywhere, and its companions, the Stone and the Cloud. They represent Lovren's homage to the Serbian Modernist poet Vasko Popa (the novel's title and the motif of an ambulatory tree are taken from his eponymous poem). This is important to note since it opens a new avenue for Serbian fantasy writers, one that offers fresh intertextual possibilities and narrative strategies. It is also characteristic of a slightly more optimistic view of the future. Much like Todorović and Nešić, Lovren strives for a communion with nature, viewing the child reader as related to nature, but not as the sole authority regarding the environment. There is no transfer of guilt and shame from the adult to the child, otherwise frequent in "green" or ecological children's books (see Bradford, 2003; Beauvais, 2015, pp. 170–178). Unlike Todorović, Lovren does not focus on humans as destroyers, emphasizing rather the healing potential embodied in Maja's efforts. It remains to be seen whether this cautious optimism will be accepted by other writers and if, in accordance with Oziewicz's hopes, the generation of our children will be able to "imagine a hopeful future—for example, the end of capitalism rather than the end of the world" (Oziewicz, 2018a).

Concluding Remarks

As we have attempted to show, Serbian folklore heritage has been adopted and used by various children's authors, who integrated these folklore motifs into their stories and fairy tales, and, more recently, in children's fantasy novels; nevertheless, recent authors lean towards other literary traditions. As the theoretical lens of ecocritical and posthumanist theory in plant-centric approach to children's fantasy literature has helped us demonstrate, the mixtures of vegetal motifs and layers of tradition in fantasy narratives encode significant statements about changes in the ecosystem and climate, ethical considerations, heritage, and futurity. Considering plant motifs therefore enriches our knowledge of the children's fantasy fiction and enhances our understanding not only of past fantasy texts but also their present and future evolution, as the narrative journey from fern flower to lilibala shows.

The findings and observations presented in this paper are limited to the corpus of Serbian fantasy fiction written for children. The next step would be to examine the significance and functions of the fantastic and magical plant motifs in other children's and YA fantasy corpora by establishing a broader Southern Slavic context or a comprehensive comparative examination of Slavic mythology in recent children's and YA fantasy.

Notes

1 Bunjak & Topić (1998) note the influence of the Polish author Adolf Dygasiński and his seminal allegory *Gody życia* on this specific fairy tale.
2 The determinants at work in the rise of Serbian's children fantasy deserve a study of their own. The most important contributing factors were as follows: a formative influence of Western fantasy writers (Tolkien and Michael Ende being probably the most popular ones among Serbian readers born in the 1960s); the breaking apart of Yugoslavia and the disappearance of the last vestiges of social

realism in Serbian children's literature; huge shifts in the publishing scene (especially towards more commercial and less overtly didactic or artistic content); and directly following the last one, the extraordinary popularity of the Harry Potter series, which convinced Serbian publishers to create a space for local fantasy novels.

3 An important exception is the gender binary: all Petrović's protagonists are male, as are the majority of the supporting characters.

4 The symbolism of this name is highly important. Namely, *Baraba* is a version of *Barabbas*, the New Testament character; however, in Serbian it is also a metonymic appelation with a strong negative connotation, meaning roughly "a rascal" or "lowlife". Petrović offers another twist and a clue to the tree's double nature when he claims that its name is actually the name of its creator in reverse.

References

Attebery, B. (2018). *From the third age to the fifth season: Confronting the Anthropocene through fantasy*. Paper presented at the Fantasy and myth in the Anthropocene international conference, Masaryk University in Brno, Czech Republic. Retrieved from https://medial.phil.muni.cz/Play/5159

Beauvais, C. (2015). *The mighty child: Time and power in children's literature*. Amsterdam and Philadelphia: John Benjamins Publishing Company.

Bradford, C. (2003). The sky is falling: Children as environmental subjects in contemporary picture books. In R. McGillis (Ed.), *Children's literature and the fin de siècle* (pp. 111–120). Westport: Praeger Publishing.

Bradford, C., Mallan, K., Stephens, J., & McCallum, R. (2008). *New world orders in contemporary children's literature utopian transformations*. Basingstoke: Palgrave Macmillan.

Bunjak, P., & Topić, M. (1998). Desanka Maksimović i Adolf Digasinjski: tri putokaza ka *Svetkovini života* [Desanka Maksimović and Adolf Dygasiński: three guideposts towards Gody zycia]. *Zbornik Matice srpske za slavistiku*, 54/55, 95–118.

Chen, P. (2018). Posthuman potential and ecological limit in future worlds. In A. Tarr & D. R. White (Eds.), *Posthumanism in young adult fiction: Finding humanity in a posthuman world* (pp. 179–195). Jackson: University Press of Mississippi.

Glavinić, G. (2016). Ekološka svest u romanu *Aven i jazopas u zemlji Vauka* Uroša Petrovića [Ecological awareness in Uroš Petrović's novel Aven and his badger-dog in the land of the Vauks]. *Detinjstvo*, 42(2), 104–114. Retrieved from http://www.komunikacija.org.rs/komunikacija/casopisi/Detinjstvo/XLII_2/12/show_download?stdlang=gb

Goga, N. (2018). Children's literature as an exercise in ecological thinking. In N. Goga, L. Guanio-Uluru, B. O. Hallås, & A. Nyrnes (Eds.), *Ecocritical perspectives on children's texts and cultures: Nordic dialogues* (pp. 57–71). Cham: Palgrave Macmillan.

Guanio-Uluru, L. (2018). Plant–human hybridity in the story world of Kubbe. In N. Goga, L. Guanio-Uluru, B. O. Hallås, & A. Nyrnes (Eds.), *Ecocritical perspectives on children's texts and cultures: Nordic dialogues* (pp. 125–140). Cham: Palgrave Macmillan.

Jaques, Z. (2015). *Children's literature and the posthuman: Animal, environment, cyborg*. New York: Routledge.

Jocić, M. (2014). Pripovedač i prirodnjak (o ilustracijama, enciklopedijaštvu, ekofantastici, ali pre svega ilustracijama Uroša Petrovića u romanu *Aven i jazopas u zemlji Vauka*) [Storyteller and naturalist: on illustration, encyclopaedias, ecofantasy, but primarily on Uroš Petrović's illustrations in his novel Aven and his badger-dog in the land of the Vauks]. *Detinjstvo*, 41(1), 70–86. Retrieved from http://www.komunikacija.org.rs/komunikacija/casopisi/Detinjstvo/XL_1/12/show_download?stdlang=gb

Jokić, J., & Marićević Balać, J. (2018). Semantika i funkcija raskovnika u *Božjim ljudima* Borisava Stankovića [The semantics and function of raskovnik in Borisav Stanković's God's People]. *Književnost i jezik*, 65(3/4), 389–400. Retrieved from https://doi.org/10.18485/kij.2018.65.3_4.6

Karadžić, V. S. (1852). *Srpski rječnik istumačen njemačkijem i latinskijem riječima* [The Serbian Dictionary, paralleled with German and Latin words]. Vienna: P.P. Armeniern. Retrieved from https://archive.org/details/srpskirjecnik00kara

Karanović, Z., & Dražić, J. (Eds.). (2016). *U početku beše reč (imenovanje biljaka u VukovomGora ljiljanova (biljni svet u tradicionalnoj kulturi Srba)* (pp. 7–19). Beograd: Udruženje folklorista Srbije: Univerzitetska biblioteka "Svetozar Marković".

Keetley, D. (2016). Introduction: Six theses on plant horror; or, Why are plants horrifying? In D. Keetley & A. Tenga (Eds.), *Plant horror: Approaches to the monstrous vegetal in fiction and film* (pp. 1–30). London: Palgrave Macmillan.

Le Guin, U. (1986). The carrier bag theory of fiction. Retrieved from https://www.deveron–projects.com/site_media/uploads/leguin.pdf

Lovren, S. (2019). *Malo zeleno drvo* [The little green tree]. Beograd: Laguna.

Maksimović, D. (1985/1935). *Ako je verovati mojoj baki* [If my grandmother is to be believed]. Beograd: BIGZ, Jedinstvo.

Milivojević, G. (1997). *Zelenselo* [Greenvillage]. Beograd: Nolit.

Nešić, I. (2013). *Zelenbabini darovi* [Greenmother's gifts]. Beograd: Kreativni centar.

Olujić, G. (2011/1979). *Sabrane bajke* [Collected fairytales]. Beograd: Učiteljski fakultet.

Opačić, Z. (2011). *Poetika bajke Grozdane Olujić* [The poetics of Grozdana Olujić's fairy tales]. Beograd: Srpska književna zadruga.

Oziewicz, M. (2018a). *Planetarianism now: Children's literature and our planet's fight for life*. Paper presented at the Children's Literature Association conference, Texas State University, San Antonio, TX. Retrieved from https://www.childlitassn.org/assets/PresenterAbstracts18/O/Marek%20Oziewicz.pdf

Oziewicz, M. (2018b). *Fantasy in the Anthropocene: Three theses*. Paper presented at the Fantasy and myth in the Anthropocene international conference, Masaryk University in Brno, Czech Republic. Retrieved from https://medial.phil.muni.cz/Play/5160

Pešikan Ljuštanović, Lj. (2009). Fantastični svet u stvaralaštvu Uroša Petrovića: između horora i humora [The fantastic world in Uroš Petrović's opus: between horror and humor]. In M. Detelić (Ed.), *Moć književnosti: in memoriam Ana Radin* [The power of literature: in memoriam Ana Radin] (pp. 55–69). Beograd: Balkanološki institut SANU.

Petrović, U. (2018). *Aven i jazopas u zemlji Vauka* [Aven and his badger-dog in the land of the Vauks]. Beograd: Laguna.

Radulović, N. (2009). *Slika sveta u srpskim narodnim bajkama* [The image of the world in Serbian folk fairy tales]. Beograd: Institut za književnost i umetnost.

Rosić, T. (1991). *Dolina jorgovana* [Lilac valley]. Beograd: Srpska književna zadruga.

Røskeland, M. (2018). Nature and becoming in a picturebook about *things that are*. In N. Goga, L. Guanio-Uluru, B. O. Hallås, & A. Nyrnes (Eds.), *Ecocritical perspectives on children's texts and cultures: Nordic dialogues* (pp. 27–40). Cham: Palgrave Macmillan.

Todorović, M. (2004). *Vir svetova* [Vortex of the worlds]. Beograd: Čigoja.

Todorović, M. (2016). *Gamiž* [Gamiž]. Beograd: Porta Libris.

Todorović, M. (2016). *Magma* [Magma]. Beograd: Porta Libris.

Tolstoj, S., & Radenković, Lj. (2001). *Slovenska mitologija: enciklopedijski rečnik* [Slavic mythology: an encyclopedic dictionary]. Beograd: Zepter.

Tropin, T. (2012). Čedožderna čudovišta ili Gamiž [Child-eating monsters, or Gamiž]. *Detinjstvo*, 38(3/4), 117–118.

Tropin, T. (2015). Preobražaji demona: motivi iz slovenske mitologije u savremenoj srpskoj književnosti za decu [Transformations of the demon: motifs from Slavic mythology in contemporary Serbian children's literature]. In D. Ajdačić (Ed.), *Slov'янsьka fantastika* (pp. 232–241). Kiïv: Osvita Ukraïni.

Velmar-Janković, S. (1998). *Knjiga za Marka* [A book for Marko]. Beograd: Stubovi kulture.

Vukmanović A. (2016). *Čudesno bilje usmene lirike* [Magic plants in oral poetry]. In Z. Karanović, & J. Dražić (Eds.), *Gora ljiljanova (biljni svet u tradicionalnoj kulturi Srba)* (pp. 69–80). Beograd: Udruženje folklorista Srbije: Univerzitetska biblioteka "Svetozar Marković".

Wolf, M. (2012). *Building imaginary worlds*. New York: Routledge.

6 Vegetal Magic

Agnieszka's Journey to the Understanding of the Vegetal Other in Naomi Novik's *Uprooted*

Mónika Rusvai

Pareidolia is a tendency to incorrectly perceive an external stimulus as an object or pattern known to the observer. A common example is seeing hidden human faces on photographs, clouds or various objects. The rugged and very often knotted surface of tree bark is also a typical source of this kind of pareidolia. Trees often get anthropomorphized through their uncanny faces staring back at us in the forest gloom. To our civilized self, these tree-bark gazes represent the untamed other. This dichotomy of self and other persists to stay at the root of our relation to nature, heavily influencing our attitudes towards the ongoing climate crisis.

Young adult speculative fiction has readily taken up the task of addressing the natural other, and in recent years the topic has been quite prolific in climate fiction. As a science fiction subgenre, climate fiction focuses on the future. The typical young adult cli-fi operates with a dystopian setting, where young protagonists are required to find science-based solutions for severe environmental damage. Putting the consequences of climate change and possible coping strategies to the forefront, it is no wonder that these novels have quickly reached a generation suffering from climate anxiety. The contemplation of possible solutions for current or future threats to civilization are undeniably important, but our understanding of nature and our relation to it are rooted in the past. Whilst science fiction attempts to provide logical solutions that science might make possible in the future, fantasy challenges us "to revise our perceptions of the natural world" (Brawley, 2014, p. 9). One of the most harmful of these perceptions is anthropocentric thinking, an attitude that can be justly blamed for our unrestricted exploitation of the natural world. However, as Ursula K. Le Guin highlights, fantasy literature is capable of escaping the trap of anthropocentrism, as it has the ability to "include the nonhuman as essential" (2007, p. 87). Contemporary young adult fantasy is rife with putting the non-human into focus: supernatural characters from sexy vampires to not-quite-deteriorating zombies have generated a wide audience. Facing the other easily fits into the coming-of-age scenarios most of these novels employ, however, the more formulaic texts in the genre hardly ever aim to challenge anthropocentric thinking through focusing on the non-human other.

DOI: 10.4324/9781032066356-9

Naomi Novik's 2016 fantasy novel, *Uprooted*, is a notable exception. Even though the novel presents a rather generic love interest, the magician called "the Dragon", Novik's focus of depicting the other lies elsewhere. Untypically, *Uprooted* leaves behind the bestial or undead others that currently invade YA fantasy, and problematizes the vegetal other. In the nameless magical wood of the novel, trees capture villagers and consume them, while the poor victims' agonized faces remain visible through the bark—an eerie explanation for the above-described pareidolia. The protagonist, Agnieszka, is a village girl living near the Wood, who is unexpectedly chosen to become the local magician's apprentice (he is either referred to as "the Dragon" or Sarkan, the latter is the first's equivalent in the magical tongue). The Dragon stands for the state-legitimized, highly anthropocentric form of magic that requires him to destroy all living beings that become corrupted by the Wood. Agnieszka is unable to internalize his perspective and begins to work out her own magic that relies on solidarity with the non-human. In my paper I examine her initiation into the wizarding (and adult) world and argue that each step of her journey leads towards a reduction of anthropocentrism, and through the blurring of the boundary between the human and the vegetal she eventually achieves a successful re-negotiation of the nature–culture divide.

Against the Wood?

Uprooted's Wood is established as the major antagonist in the very first paragraph, when Agnieszka claims that the Dragon protects the village "against the Wood" (Novik, 2016, p. 3). At this point, the Dragon's position is left ambiguous, the only certain element being that he is not allied with the common enemy: "He wasn't evil, but he was distant and terrible" (Novik, 2016, p. 10). Agnieszka is telling her story in the first person, and at first, she completely identifies with her community and adopts their view of seeing the Dragon as the lesser evil. If we look at this situation through Farah Mendlesohn's system for categorizing fantasy, we can easily observe that the focal character is immersed in her world (the existence of magicians and witches is accepted) and in this system the Wood's corrupted magic is an intruding force (2008). Exactly as Mendlesohn highlights when describing intrusion fantasy, the interaction with the landscape in *Uprooted* is motivated by fear instead of wonder. Relying on this intrusion fantasy structure, Novik establishes the opposition of nature and culture early on in the novel.

This opposition, however, is never realized in the way it is depicted on the Dragon's map. On this map, pinned on the wall in the tower's topmost room, Agnieszka perceives a clearly set boundary between woodland and human civilization. In reality, however, the Wood's margin is not a demarcated border, but an inverse-frontier: nature stretching its leafy hands to consume the human world through magical corruption. It already destroyed a village a generation before. Interestingly, though, looking at this map also serves as the first moment of detachment for Agnieszka: briefly adopting the

Dragon's perspective, despite their separation, villages and Wood become a single entity in relation to the rest of Polnya. Villagers in the Wood's valley never move from their immediate homeland, whilst other Polnyans are afraid to enter the Wood's vicinity. The confrontation of man and nature is thus encapsulated in this place, making the Dragon's valley fairly close to Stefan Ekman's term, the *polder*. A polder is a special area within a fictional land that involves "maintained anachronism" (Ekman, 2013, p. 118). The Wood in *Uprooted* does not directly intervene into the course of time, but people struggling with its malevolence are clearly isolated from the rest of civilization, resulting in a different, local history, that takes no notice of political incidents outside the valley. The only representation of the state within the valley would be Sarkan, the state magician, but apart from his occasional help against the Wood, villagers are left to their own devices in coping with the ever-changing nature–culture boundary.

Agnieszka only gains a deeper insight into this picture once she is unexpectedly chosen to be the Dragon's new apprentice. In a matter of weeks, she has to face the fact that the unfamiliar magic she fears in the Dragon is something she herself also possesses, and she is taken to receive training as a witch. However, it soon turns out that the magic Agnieszka has unknowingly possessed since early childhood has always been linked to nature. "You'd go into the forest and come back with fruit out of season, or flowers no one else had ever seen. When we were little, you always used to tell me stories the pines told you", says Agnieszka's best friend, Kasia, when the two of them attempt to trace early signs of Agnieszka's magical skills (Novik, 2016, p. 65). This realization already sows the seed of thinking in less hostile terms about the Wood's magic. The vegetal aspect of Agnieszka's magic becomes a major intruding force in her life, creating a small trajectory of intrusion within a larger framework, following the "from denial to acceptance" route in the Mendelsohnian sense. A closer look at Agnieszka's learning process does not only support such a trajectory, it also shares the characteristics of female coming-of-age narratives. As Brian Attebery observes, women's coming-of-age stories are focused on unleashing rather than mastering power (1992, pp. 88–89), and this is exactly what happens in *Uprooted*. Agnieszka's first experiences of the Dragon's magic are fairly similar to her confrontation with the map of the valley: she understands that there are signs, but she is yet unable to understand them. The same thing happens when she steals a book from the Dragon's library, and no matter how hard she tries, understanding deludes her every time, even though she is able to read it (Novik, 2016, p. 27). Thus, to unleash her skills, Agnieszka has to gain a good command of the Wood's vegetal magic that feeds her power.

Unfortunately, the road paved by the previous apprentices of the Dragon is not one Agnieszka could readily follow. All spells feel "wrong" on her tongue, she seems unable to control her magic through the prescribed mechanics. Even though Novik invokes the trope of the chosen one at the beginning, Agnieszka is not a heroine to go down the well-trodden path of destiny, her route is closer to what Mendlesohn identifies as the one of the intrusion fantasy hero(ine): a person who challenges the established system of knowledge and evidently rewrites its rules. Interestingly, in the first case when Agnieszka modifies a spell, she is prompted by Sarkan, but once she recognizes that any such language modification equals a much more efficient agency over magic, she goes on experimenting wildly, as she already knows that she can only aid her family and friends against the Wood through magic.

It gradually becomes clear to Agnieszka that she and Sarkan are using two radically different methods to control their magic. Sarkan's magical practice is governed by detailed, developed and proven rules. This brings his methodology fairly close to our modern-day approach to natural sciences even with respect to social status: it is the hegemonic practice of state magicians and witches, clearly accepted by Sarkan as default. His spells are all premediated, every modification strictly calculated and working towards a previously identified goal. For him, words of the magical language have a set meaning, and language as a system relies on strict rules—much like the magic it induces. As opposed to this, Agnieszka's handling of magic openly relies on intuition-based modification of language, and the focus from a clearly set goal shifts to emphasize the importance of the journey. After she heals Sarkan's infected wound (her first major achievement using magic her own way) she compares magical practice to finding your way in the wood: "You have to pick your way through the thickets and the trees, and it's different every time" (Novik, 2016, p. 88). Whilst Sarkan would readily struggle his way down the road of the patriarchal hero monomyth towards the inherently righteous aim of defeating the antagonistic force, Agnieszka would stop to smell the flowers, and repeatedly ponder the justness of her cause and the nature of her enemy.

These differences in practical magic become even more apparent through the lenses of the nature–culture dichotomy. Sarkan's method seems to rely on Timothy Morton's concept of the Severing: "a foundational, traumatic fissure between, to put it in stark Lacanian terms, *reality* (the human correlated world) and *the real* (ecological symbiosis of human and nonhuman parts of the biosphere)" (2017, p. 13). In addition to this separation from nature, the Severing also involves a "privatization of access to the real" (2017, p. 20)—in the world of *Uprooted* this means limiting the access to magic to state-appointed magicians. Hegemonic magic in Polnya is language-based and thus inherently political, cultural and anthropocentric. As Morton observes, a young adult is basically "an anthropocentrist in training" (2017, p. 15), and Agnieszka as Sarkan's apprentice receives exactly this kind of training. But she resists and with each step of resistance she gets ever deeper into the imaginary wood of her own magic practice. Being in the Wood provides a detached position, and new perspective on magic. Based on her above-quoted observation that the metaphoric forest of magic is different every time, Agnieszka is ready to modify the human-made rules of hegemonic magic according to her needs. Going down this road she discovers that meaning should not necessarily be restricted to anthropocentric semiotics; and rhythm, intonation and even the amount of magic she feeds into her spells have their own meaning.

As opposed to Sarkan's "laboratory magic" (he literally has a laboratory in his tower) that is expected to be perfectly calculable, Agnieszka instinctively refrains from all magic that resembles a programmed mechanism. Her creativity in modifying spells is not only similar to the inherent incalculability of nature, itis actually fuelled by it both in the physical sense (through the application of herbs) and through sympathetic magic. Despite her first experience of detachment when she sees the map of the valley on the wall, Agnieszka is still deeply rooted in nature through her magic, making it questionable whether the Wood is her real antagonist.

Bodies Transmuted

On the surface, Novik's method of picturing the Wood as the fearful vegetal other is rather conventional. Relying on all five senses, she describes how corruption leads to the malformation of crops and livestock, she makes her reader smell the pollen, touch the sap from the trees and listen to the ominous rustling of heart tree leaves. Entering the Wood entails a feeling of claustrophobia, a fatal detachment from the human-regulated world. Yet the major source of horror is the supernatural acceleration of vegetal growth. This method clearly echoes Dawn Keetley's third thesis of plant horror: "plants menace with their wild, purposeless growth" (2016, p. 13), and it is probably among the most common ways of plant modification in fantasy literature. In Novik's Wood, tree bark grows so quickly that it covers an entire human body in a matter of seconds, vines are fast-moving arms reaching towards their enemies and heart trees grow with unimaginable speed. All this unbounded growth seems to culminate in the description of the Wood-queen: "There was too much of her. She could have filled in any holes we made in the world and still had more of herself left over" (Novik, 2016, p. 386).

The only reason why this monstrous growth still remains contained, is the fact that vegetal growth does have a purpose in *Uprooted*. All vegetal beings are highly anthropomorphized in Novik's vision, hence walkers, mantises, corrupted crop and livestock, beasts and birds of the forest, heart trees and even pollen become agents of the Wood's undecipherable intentions. Even though anthropomorphic thinking is approached with a generous amount of scepticism (if not open hostility) in contemporary ecocriticism, *Uprooted* manages to remain eco-conscious. Firstly, the naïve anthropo*morphic* thinking of villagers helps to blur the distinction between the human and vegetal hegemonic magic has been struggling to maintain. Secondly, the state-enforced anthropocentric thinking that would require the destruction of the Wood in the name of human safety is challenged by the introduction of a human-vegetal hybrid viewpoint manifested in the transmuted bodies of Kasia and Queen Hanna.

Both Kasia and the queen end up inside heart trees (Kasia is taken there by walkers, the queen's exact story remains unknown), but once they are released from the physical grasp of vegetation, it turns out that they had been consumed from the inside all through their captivity. Less serious cases of corruption also start with some piece of the Wood getting inside the victim's body: eating corrupted food, inhaling pollen, being bitten by a corrupted beast. When someone is taken by a heart tree there is a reversal of inside and outside, but the result is the same—the victim becomes irreversibly infused with the vegetal. Hybridity thus achieved defies categorization within the hegemonic magic system, turning all such creatures into monsters: the third thesis of Jeffrey Jerome Cohen's monster theory assumes that a monster is necessarily a "harbinger of category crisis" (1996, p. 6). Fortunately for the corrupted, Agnieszka adopts a similar approach to monstrosity. As Cohen reminds us, the *monstrum* is etymologically "that which reveals", "that which warns" (1996, p. 4)—similarly, in Novik's text, the vegetal-human monster is evidently a sign, an initially indecipherable magic spell that Agnieszka makes repeated efforts to understand.

On this quest for meaning, Agnieszka is clearly urged by the abject quality of monstrous hybridity. Adopting hegemonic thinking would mean that she has to kill her corrupted best friend, Kasia, but seeing her in a human shape being consumed by the Wood from the inside leaves the young witch no choice but trying to save her. Although *Uprooted* provides a highly anthropomorphized picture of the vegetal, this understanding of nature is presented as a useful tool for villagers of the valley to cope with alterity. Their anthropomorphizing approach is very much in line with Jane Bennett's assumption that anthropomorphism might prove a useful tool against anthropocentrism (2010, p. 120)—in our case against the overwhelming power of hegemonic magic. While state magicians persist in the extermination of all corrupted beings based on their own anthropocentric interest, village-trained Agnieszka is ready to perceive both the human features in the vegetal (for instance the tree-bark faces) and the vegetal within the corrupted humans. This perspective leads her to the key realization that the trees themselves are also corrupted, and the agency of an unknown evil is imposed upon them (Novik, 2016, p. 125). From this point onwards, Novik's rhetoric is getting ever clearer about the agency of the entity behind the Wood: "a living thing deliberately reaching out the full force of its malice" (2016, p. 181) and then "[t]he Wood had set us one trap after another, over and over" (2016, p. 285).

Thus, the previously enforced dichotomy of thinking is broken. In order to save her homeland, Agnieszka has to track a chain of causality—a mysterious entity using the Wood to destroy the valley—in a new system in which the inter-relatedness of living beings is far more complicated than the calculable mechanisms of Sarkan's magic would have her believe. Even though she manages to purify Kasia, the Wood leaves her body "transmuted and changed" (Novik, 2016, p. 147). Her limbs become rigid; her face turns so beautiful it is as if it was carven from wood, yet the spirit inhabiting her body remains the same. Nevertheless, her uncorrupted otherness is unacceptable for the magical and political leaders of Polnya. Queen Hanna, however, complicates the issues of inter-relatedness and bodily construction even further. Once taken to the Wood by a Rosyan prince, the queen spent the last 20 years within a heart tree being consumed by the Wood from the inside. When she is rescued, Agnieszka's and Sarkan's spell shows no trace of corruption in her—but nothing else either. What is left of the queen is no more than a walking wooden shell without soul. In a sense, she has literally become a corpse, re-introduced into the cycle of nature: the tree fed on her body and turned it into its own material. Agnieszka, whose magic relies on situation-based modification, can train herself to face the uncorrupted otherness of Kasia, but the queen's bodily construction proves too much of a stretch for her—at first.

Magic as the Vitality of Matter

The plant kingdom blurs another important category boundary: the one between life and death. How would you decide if a tree is dead? If you stuck the branch of a fallen willow tree into the earth it might burst into new life, and we are getting ever closer to scientific evidence for tree trunks being kept alive through a system of roots by neighbouring trees.[1] In *Uprooted* this phenomenon is magnified through the lenses of the fantastic, and the reader is informed early on that all corrupted creatures are hard to kill if they can be killed at all (Novik, 2016, p. 68). The queen may be dead in spirit, but her transmuted wooden body has preserved its vitality and is able to move about even without a human consciousness behind it.

During Agnieszka's stay in Kralia (the capital of Polnya modelled on medieval Krakow)[2] the question of life, death and agency looms in the background, as the young witch tries to figure out whether the queen's altered body is still in the grasp of the Wood. Her discoveries complicate the inter-relatedness of human and non-human even further. Firstly, she is yet again confronted with the fact that magical people have a far longer life span than others (she learns that Sarkan himself is around 150 years old), which means that with respect to time perception, these people are actually closer to trees than to humans. Later on, the state magicians reveal the reasons why Sarkan took apprentices from Agnieszka's valley in the first place: "There is some power deep in your valley, some strangeness beyond mortal magic that draws men in, plants roots in them—and not only men" (Novik, 2016, p. 281). The strange power previously attributed to the Wood is now extended to the entire valley, turning it into a superorganism tied together with invisible roots of magic. To some extent, this would explain the vitality of the queen, but Novik challenges her reader to stretch his concepts even further: the Wood's power is expanded to objects when the corrupted bestiary from the library wreaks havoc on the royal palace.

Extension of agency to a non-human, non-living other through magic is not unknown in fantasy literature. Calling attention to this great interrelatedness of all organic and inorganic matter is fairly close to what Jane Bennett describes as "a shared, vital materiality" (2010, p. 14). Bennett argues for the agential roles of various organic and inorganic materials around us (including the food we consume), but highlights the fact that this "Thing-Power" is not necessarily linked to a consciousness: "[Thing-Power is] the curious ability of inanimate things to animate, to act, to produce effects dramatic and subtle" (2010, p. 6). Her theory lays down the foundations of a non-anthropocentric approach towards organic and inorganic matter. In her view, our reconciliation with the materiality of the world is rooted in the acceptance of our own foreignness, since human bodies involve both human and non-human components, in our material reality, we ourselves are "an array of bodies" (2010, pp. 112–113). In the real world, however, such knowledge either remains suppressed, or consciously buried under the hegemonic narrative of anthropocentrism. The magic lurking in Agnieszka's valley is a possible equivalent to Bennett's vitality of matter, a power that lays bare the foundations of our very being: the non-human other rooted within the human self. As Rosemary Jackson points out, this reconciliation of self and other is an essential feature of the fantastic: "One of the central thrusts of the fantastic is an attempt to erase this distinction itself, to resist separation and difference, to re-discover a unity of self and other" (1981, p. 30).

By the time Agnieszka returns to her homeland, she is already aware of this invisible system of magic linking the living and the non-living from the valley. She also knows that the Wood-queen stands at the end of the causality chain, and the only way to challenge her is to use the same magic system that she relies on. Now Agnieszka openly manipulates everything Sarkan taught her to suit her own needs: she uses even more natural materials, such as herbs, even soil, walks barefoot to let the power of the valley flow inside her, and cultivates her sympathetic magic to perfection. When she finally re-enters the Wood to find the escaped queen, the two of them resemble each other more closely than ever before.

However trained Agnieszka is in the ways of the Wood, a significant barrier between her and the Wood-queen remains. Except for the maddened snarling and swearing of corrupted people, the Wood in the queen's hold is incapable of verbal communication. Yet during the siege of the Dragon's tower it becomes clear that the possibility of defeating the Wood-queen is closely linked to her personal history. Knowledge of the past—from our anthropocentric view at least—is so closely intertwined with story, that it is hard to imagine any narrative in absence of language. Should we accept then, what Luce Irigaray concludes about the language of the vegetal, that a tree teaches us in silence (2017, p. 129)? But how is this silence that hides vegetal meaning constructed? In Irigaray's view plants "say through shaping their own matter" (2017, p. 129). Consequently, the silent vegetal narrative of heart trees equals their own materiality, a materiality that in *Uprooted*'s world is infused with magic: every creature that has ever tasted water or food from the valley becomes metaphorically rooted in the place. That is the reason why people living near the Wood persist to stay despite the obvious danger.

As Agnieszka is now able to perceive this vital materiality encompassing everything in the valley, she is ready to become part of its super-organic system to learn the origins of the Wood-queen. Once taken by a heart tree, she goes through the classic death-and-rebirth scenario of the hero monomyth, and she returns with the recognition that the boundaries of vegetal and human have always been blurred within her very body:

> I might have been the daughter she'd [the Wood-queen] hoped for, the child halfway between the tower-people and her own; she might have been my teacher and guide, like Jaga's book showing me the way. We might never have been enemies at all.
>
> (Novik, 2016, p. 420)

When later on Agnieszka returns to the Dragon's tower, she finds the map of the valley torn, its magic gone (Novik, 2016, p. 422). Even though she assumes that the map must have been destroyed during the siege, Agnieszka knows that the hold of hegemonic magic is broken over the valley. Even Sarkan has submitted to the power of the Wood: during their final battle against the Wood-queen he drinks from the forest stream to be able to save Agnieszka, hence his power is now intertwined with the magic of the valley. Magic is now literally (and literarily) operating as vitality of matter: "an active becoming, a creative not-quite-human force capable of producing the new buzzes within the history of the term nature" (Bennett, 2010, p. 118).

Even though Agnieszka's recognition of the inter-relatedness of the human and the non-human induces a slow change in people's perception of nature, her achievement remains geographically limited to the valley, and as a wandering witch of the woodlands she is unlikely to ever get re-integrated into everyday society. Thus, *Uprooted* challenges the formerly unquestioned anthropocentrism of the hegemonic magic system, but following the hero's journey structure, it refrains from offering a solution extended to the entire society. Change in fantasy fiction—as many times in real life—is often initiated by an extraordinary individual, and it only, if ever, trickles down to society later. Neither is the hero's path limited to a single meaning. In *Uprooted*, Novik manages to intertwine the generic coming-of-age plotline of a wizard's apprentice with a serious contemplation of the dangers of anthropocentric thinking—the former is clearly visible, the latter is lurking in the shadow of the trees waiting for the well-trained reader's eyes. This haunting theme wrapped in a popular scenario might be key to why such an eco-conscious text remains attractive and accessible for young readers.

Notes

1 Scientists have suspected for a long time that a sharing network exists among trees in the forest that is capable of transmitting water and nutrients. In 2019, researchers found a kauri tree stump in New Zealand that was apparently kept alive by such a system of roots and started to look further into this strange phenomenon (Mayes, 2019). If their presumptions are proven, it would mean a great step towards scientific evidence that forest ecosystems work as superorganisms.

2 As Novik explains in one of her interviews, she based Kralia on medieval Krakow, as she wanted to re-create her childhood image of Poland: "a world slightly blurry around the edges, like a watercolour" (Interview with Naomi Novik, n.d.).

References

Attebery, B. (1992). *Strategies of fantasy*. Bloomington: Indiana University Press.

Bennett, J. (2010). *Vibrant matter: A political ecology of things*. Durham, NC: Duke University Press.

Brawley, C. (2014). *Nature and the numinous in mythopoeic fantasy literature*. Jefferson, NC: McFarland & Company.

Cohen, J. J. (1996). Monster culture (seven theses). In J. J. Cohen (Ed.), *Monster theory: Reading culture* (pp. 3–25). London: University of Minnesota Press.

Ekman, S. (2013). *Here be dragons: Exploring fantasy maps and settings*. Middletown, CT: Wesleyan University Press.

Interview with Naomi Novik. (n.d.). Retrieved from: https://www.portersquarebooks.com/interview-naomi-novik

Irigaray, L. (2017). What the vegetal world says to us. In M. Gagliano, J. Ryan, P. Vieira (Eds.), *The language of plants: Science, philosophy, literature* (pp. 126–135). Minneapolis: University of Minnesota Press.

Jackson, R. (1981). *Fantasy: The literature of subversion*. London: Methuen.

Keetley, D. (2016). *Plant horror: Approaches to the monstrous vegetal in fiction and film*. Basingstoke: Palgrave Macmillan.

Le Guin, U. K. (2007). The critics, the monsters, and the fantasists. *The Wordsworth Circle*, 38(1/2), 83–87. https://doi.org/10.1086/TWC24043962

Mayes, K. (2019). Trees share water to keep this dying stump alive. Retrieved from https://www.sciencemag.org/news/2019/07/trees-share-water-keep-dying-stump-alive.

Mendlesohn, F. (2008). *Rhetorics of fantasy* [ebook version]. Middletown, CT: Wesleyan University Press.

Morton, T. (2017). *Humankind: Solidarity with non-human people*. London: Verso.

Novik, N. (2016). *Uprooted*. London: Pan Books.

Part III

Arboreal Embraces

7 Arboreal and Maternal Desires

Trees and Mothers in recent Australian Middle-Grade Fiction

Melanie Duckworth

When Molly's mother accidentally turns herself into a tree, in Martine Murray's *Molly and Pim and the Millions of Stars* (2015), she invokes a long tradition linking the maternal with the arboreal. From Ovid's Myhrra, who gives birth while in tree form (2004/AD 8, p. 358), to biologist Suzanne Simard's description of "mother trees" that nurture and fortify younger trees in forest ecosystems (2021); from St Bonaventure's 13th-century interpretation of the virgin Mary as the Tree of Life (2011), to Shel Silverstein's unsettling portrayal of maternal-arboreal self-sacrifice in his picturebook *The Giving Tree* (1964/2014); mothers have been associated with trees. Trees and mothers tend to be regarded in the context of what they can provide for others: trees provide shade, fruit, and wood for fuel and shelter, while motherhood is often understood as abrogating one's own desires in favour of the needs of one's offspring. The desires *of* trees and mothers are less frequently considered. While the concept of "arboreal desire" depends upon a degree of anthropomorphism, reflecting on the "desires" of mothers and trees facilitates a focus on entities traditionally backgrounded in literature for children.

This chapter examines the representation, and conflation, of trees and mothers in three recent Australian middle-grade novels: Murray's *Molly and Pim and the Millions of Stars, The Mulberry Tree* (2018), by Allison Rushby, and *How to Bee* (2018), by Bren MacDibble. In *Molly and Pim*, Molly's mother literally becomes a tree, and in *The Mulberry Tree*, an ancient tree's grief over the loss of her "daughter" prompts her to prey upon the human daughters of the village. In MacDibble's dystopian *How to Bee*, where children work as "bees" to pollinate trees in a world where wild bees are extinct, trees provide refuge to Peony, who is threatened by her mother's reliance on the city and an abusive relationship. These books were chosen as they each both invoke and trouble the trope of mother as tree—nurturing and self-sacrificial—that has been embodied disturbingly in Silverstein's *The Giving Tree*. The various constellations of tree–mother–child relationships in these novels interrogate themes of kinship, nurture and maternal desire.

Drawing on ecofeminism, critical plant studies, and Donna Haraway's notion of "making kin", I examine the ways in which the desires of mothers

DOI: 10.4324/9781032066356-11

and trees are juxtaposed, merged, and obscured in the novels. The representations of trees in these novels can be linked to an increasing awareness of the global environmental crisis, as well as, in some cases, a knowledge of scientific discoveries about plant sentience and communication. Each novel, in a different way, advocates a close relationship with trees and hints at an understanding of trees as subjects with their own needs and reasons for being. The mother figures in these texts at times confirm but at other times resist the image of a mother as a nurturing, steadfast tree. Positioned in matrices of arboreal and maternal desires, the child characters learn to navigate worlds in which, as Jane Bennett puts it: "human culture is inextricably enmeshed with vibrant, non-human agencies" (2010, p. 108). Ultimately, trees and mothers are more than metaphors, but the powerful image of the mother tree influences their narrative potential and colours the ways in which they are perceived.

Mother Trees

The notion of the mother tree is linked to the broader conceptual category of Mother Earth. The physical processes of pregnancy, childbirth, and nursing are some of the reasons women have been conceptually linked with the "natural", the "animal", and even the "botanical". As Anna Katrina Gutierrez notes,

> We are able to make sense of the integrated ideas Mother Earth and Mother Nature by identifying similarities in the vocabulary associated with schemas of motherhood and nature/Earth, such as 'nurture', 'nourish', 'fertile', and fecund,' and in vital relationships, such as 'source of food, water and shelter' and 'treat with love and respect'.
> (2016, p. 133)

This "natural" mapping of the arboreal and the natural onto the maternal has been critiqued by ecofeminists, who see a link between the oppression of women and the oppression of the earth (Plumwood, 1993; Merchant, 1980). Ellen Cronan Rose argues that the very concept of "mother earth" is problematic, and potentially damaging to both women and the natural world (1991, p. 91). Nevertheless, the association is pervasive. Women and girls are often associated with flowers, the decorative sexual organs of plants (Hines, 2004; Jackson-Houlston, 2006). And, as Anissa Janine Wardi points out, "trees are a particularly maternalized plant" (2016, p. 31): "In so far as the discourse of kinship lines is mapped onto trees, (i.e. tree of life and the family tree) trees are simultaneously maternalized and rendered as sites of history and memory" (Wardi, 2016, p. 30). Zoe Jaques notes that as both "durable" and "fragile", trees can help us reflect on our own mortality. Discussing a short story in which the author describes being comforted by a willow tree bending over her mother's grave, Jaques draws attention to "a cultural ethos that ... locates trees as venerable and sentient watchers over the succession of human generations", operating as "silent, kindly cipher[s] for nature's paradoxical permanence" (2015, p. 112). This sentiment is also found in fairy tales. In the version of Cinderella told by the Brothers Grimm, there is no fairy godmother, but Cinderella is helped by a magical hazel tree growing on her mother's grave (1972, p. 122). In both of these stories, a tree stands in the place of a mother, responding to a child character's desire for maternal care and protection.

As well as being potent emblems of lost mothers, trees and other plants provide care for humans and animals by providing the oxygen we need to breathe. Luce Irigaray imagines her relationship with the vegetal world as analogous to the relationship between an unborn baby and its mother:

> Perhaps I could say that the vegetal world had become a mothering place that provided me with the air I needed. However, I was no longer a fetus. I no longer received oxygen with the maternal blood through the umbilical cord and the mediation of a placenta. I was already born and had to breathe by myself. It was the vegetal world that ensured mothering care with the environment it arranged around me.
>
> (2016, p. 21)

This image of a child relishing in the care and protection of the vegetal world is a theme often taken up by children's picturebooks, as Casals and Mayne-Nicholls, and Goga have shown in their chapters in this volume. A similar care and abandon in a joyful relationship between a tree and a child is evident in the opening pages of Silverstein's *The Giving Tree*, in which a boy "loved the tree very much. And the tree was happy" (1964/2014, unpaginated).

The Giving Tree has been read as both a fable of environmental destruction and a discomforting depiction of sacrificial—and co-dependent—motherly love. In this book, the desire of the tree—who can be read as a mother—is for the happiness and the proximity of the boy. In the course of the narrative, he grows into a man and exploits her willingness to please him—he eats her fruit, removes her branches, cuts her down, and ultimately rests on her empty stump. As Greta Gaard puts it: "the tree stands in for the all sacrificing mother in a hetero-patriarchal culture, and the narrative constructs humanity as a narcissistic little boy who keeps asking more and more of nature (the tree) until there is nothing left" (2009, p. 16). In this story, the desires of the boy and the tree are equally problematic: the boy wants too much, and the tree wants to please the boy so much that she is happy to destroy herself. This mapping of maternal, self-sacrificial desire onto a tree is troubling on two levels. Firstly, it depicts an oversimplified and pathological version of the maternal, and secondly, it presents a dissonant and anthropomorphized picture of arboreal desire. If trees can be said to desire at all, their desire is surely not for the physical closeness of a human being at the expense of their own destruction.

Arboreal Desire

To import a human emotion—desire—to a vegetal being is problematic. Fruit trees have long been regarded as emblematic of desire, as the tree of knowledge in the Garden of Eden attests (Genesis 3:6). Fruit is sweet in order to entice animals to eat it, as Michael Pollan points out: "Desire, then, is built into the very nature and purpose of fruit, and so, quite often, is a form of taboo" (2001, p. 19). It is questionable, however, whether trees themselves can be regarded as desiring beings. Natania Meeker and Antónia Szabari regard the concept of desire, transposed to plants, as entirely inappropriate, and affirm the French botanist Francis Halle's view that "animal properties such as interiority, desire, mortality and individuality are largely meaningless for understanding plant life" (2020, chapter 1, paras 13–14). The profound indifference of plants to our own desires for knowledge, beauty and relationships impedes a connection with plants on these terms, and Meeker and Szabari stress "the limitations of a model of plant-human interaction based on desire or language or indeed any recognizable human qualities" (2020, chapter 1, para 11). Apparently lacking subjectivity, plants surely lack desire.

On the other hand, as Michael Marder points out, in the *Timaeus*, Plato describes plants as possessing an "appetitive soul", "the same as the one housed in humans between the midriff and the navel. This soul … 'shares not at all in opinions and reasoning and mind but sensation, pleasant and painful, together with desires"' (Marder, 2014, p. 12). According to Plato, humans have a tripartite soul encompassing appetite, spirit and reason, whereas plants are restricted to appetite. The appetitive soul is not reflective or self-aware, but immediate and sensual. A plant, which needs water, wilts when it does not get enough, and thus can be said to "desire" water. As Marder puts it, "If they can be 'thirsty', and if desire is associated with the experienced lack of the desired thing in the desiring being, then the fern you have not watered for weeks is, in fact, desirous for water" (2014, p. 13). In this sense, we can speak of plants desiring water, light and perhaps even the means of reproduction. The poet Kahlil Gibran writes that "it is also the pleasure of the flower to yield its honey to the bee" (1923/2020). This notion that the blossom desires the bee is not merely sympathetic fallacy. The tree *needs* the bee to pollinate its flowers, and thus can be understood, in some senses of the word, to "desire" not only life itself but also the means to reproduce: the flower, the bee, the fruit.

In *The Botany of Desire*, instead of viewing domesticated plants as passive entities cultivated, controlled, and selected by humans, Pollan regards plants as active beings that manipulate the desires of humans and other animals for their own gain. He sees wheat and cotton as some of the most successful plants on the planet, as they have co-opted humans into clearing land for them and cultivating them on an enormous scale. He writes that plants and humans "have been actors in co-evolutionary drama, a dance of human and plant desire that has left neither the plants nor the people taking part in it unchanged" (2001, p. 243). Pollan focuses his analysis on four abstract qualities that humans desire and receive from plants: sweetness, beauty, intoxication and control. Only on rare occasions does he mention plants' own desires, but the understanding that plants are active subjects underpins his book: "desire might be too strong a word for whatever it is that drives plants to reinvent themselves so that we might do their bidding, but then, our own designs have often been no more willful than the plants'" (2001, p. 243). Perhaps desire is "too strong a word", but by writing "desire", not merely "drive", Pollan envisages plants as active beings, while at the same time questioning the freedom and individuality that humans assume underpin the pursuits of their own desires.

Maternal Desire

In *Maternal Desire*, Daphne de Marneffe claims that the concept of "desire" tends to be regarded as incompatible with motherhood: "[h]istorically, *mother* and *desire* do not belong in the same phrase. Desire is about sex, and motherhood is about practically everything but sex" (2019/2004, pp. 20–21). She argues that while feminism has rightly focused on achieving equal rights for women, including freedom from the constraints of motherhood, the movement has tended to overlook and delegitimize maternal desire. Marneffe reclaims the word "desire" from its associations with sex and argues that the maternal desire to nurture is positive and liberating:

> The desire to mother is not only the desire to have children, but also the desire to care for them … It is the longing felt by a mother to nurture her children; the wish to participate in their mutual relationship; and the choice, insofar as it is possible, to put that desire into practice.
>
> (2019/2004, p. 9)

"Maternal desire", however, is never really a singular thing, and the desires of mothers are often contradictory. Ambivalence is, Barbara Almond argues, a ubiquitous experience of motherhood (2010, chapter 1). Maternal desire can involve feelings of love and nurture, but also fear, anger and vengefulness. Mothers desire many things—to be with their children and to be away from them, work, relationships, freedom. It is thus more accurate to speak of a plurality of maternal desires, some of which map onto the metaphor of a tree better than others.

In children's literature, maternal desires are generally (and understandably) seen as secondary to the character development of child protagonists. The characters of mothers are frequently required to get out of the way of the narrative in order for the child to experience her own adventure. The desire of the mother to protect and nurture can stifle a child's narrative potential. This may account for the many orphans in children's literature, as can be seen for instance in J. K. Rowling's *Harry Potter* series (1997–2007), Francis Hodgson Burnett's *The Secret Garden* (1911/1998), and J. M. Barrie's *Peter Pan* (1911/2015). On the other hand, As Lisa Rowe Faustino and Karen Coats point out, "[the mother's] guises in the literature written for children seem almost cartoonish given the complexity of her position within a child's actual life" (Fraustino & Coats, 2016, p. 3). By paying attention to the desires of mothers depicted in the texts, we can get a clearer picture of the complexity of their roles. Marneffe defines "maternal desire" as "a desire anchored in [a mother's] experience as an agent, an autonomous individual, a person" (2004/2019, p. 10). Considering "plant desire" achieves a similar goal. Matthew Hall regards plant sensing, plant communication, plant intelligence and plant self-awareness as evidence that we should consider plants as persons (2011). Regarding trees and mothers as desiring individuals rather than just objects of desire, enables us to approach them as persons and actors.

To speak of a reciprocal "dance of human and plant desire" (Pollan 2001, p. 243) is one way of seeing plants as kin. In *Staying with the Trouble*, Haraway invites her readers "to make [non-human] kin in lines of inventive connection as a practice of learning to live and die well with each other in a thick present" (2016, p. 1). Desire plays a role in the process of making kin. An orchid shaped like an extinct bee "collects up the presence of the bee aslant, in desire and mortality" (Haraway, 2016, p. 69). Haraway's catch cry to "make kin, not babies", however, predicates a transformative and healing relationship with the natural world with the erasure—or rechannelling—of maternal desire (Haraway, 2016, pp. 5–6). *Molly and Pim, The Mulberry Tree*, and *How to Bee* engage to varying extents with a perceived need to "make kin" of trees, and to live more harmoniously with the natural environment, in the context of a global shift to greater awareness of environmental concerns. *How to Bee* makes this goal explicit as it is set in the aftermath of an ecological catastrophe, *The Mulberry Tree* invokes notions of plant sentience and "mother trees" in a meditation on vengeance and forgiveness, and *Molly and Pim* draws on pagan traditions of herbal magic in order to emphasize the inter-relatedness of botanical and cultural worlds.[1]

Molly and Pim and the Millions of Stars

In *Molly and Pim*, Molly's mother desires and partakes in a rich and unusual relationship with plants. She is a vegetarian herbalist who wakes early to gather herbs while their vibrations are strongest, and turns them into magical potions. Molly is embarrassed by her unusual mother and their unique lifestyle: "Molly didn't like to think about plants vibrating or emitting or sensing, as this was all part of the strangeness of things, and she strongly objected to strangeness and tried to pretend it wasn't there" (Murray, 2015, p. 2). Thus Molly's mother's desire for plants complicates—or at least, shapes—her equally strong desire to nurture Molly, which is evident in the way she appears "so relieved to see Molly that she glided out of the bush as if the task of gathering Molly up and steering her home suddenly gave her the right weight in the world" (p. 19). When a conflict with their grumpy and strait-laced neighbours escalates, Molly's mother's solution is to grow a huge, beautiful tree to restore harmony and block them out. Molly points out that a tree usually takes a long time to grow, but her mother replies: "Yes, usually, but I think I can make it grow very quickly" (pp. 28–29). Unfortunately, Molly's mother accidentally drinks the powerful herbal potion she has prepared to speed the growth of the acorn. Realizing her mistake, she "looks at Molly with great yearning", before transforming into a trunk, roots and branches (p. 52).

The description of her transformation is reminiscent of similar events in Ovid's *Metamorphoses*. Ovid's metamorphosed trees begin frequently (though not exclusively) as women fleeing unwanted sexual advances or other dangers. Their tree form protects them. One of his characters, Myrrha, is pregnant at the time of her transformation, subsequently gives birth to Adonis while in the form of a tree (Ovid, 2004/AD 8, p. 358). Hall interprets this "interpenetration of human and plant at birth" as evidence that "human beings and plants are displayed as having a deep and substantial kinship connection that transcends their temporary forms" (2019, p. 72). Unlike Ovid's characters, the transformation of Molly's mother is not a form of protection, but an apparent accident. The accident is made possible by a number of contributing factors: her desire to live in harmony with the vegetal world, her extensive knowledge of the potency of vegetal magic, and her weariness from working all night on the potion. It also attests to the connectedness of things. As Molly and her friend Pim contemplate how to change her back, Pim observes:

> I think that the kind of magic we need is going to come from being close to things. I mean trees, animals, earth, sky. We're all made of dirt or blood or sap and air and spirit and stuff. We are all sort of the same. (pp. 184–185)

The metamorphosis can take place because humans and trees are kin.

As a tree, Molly's mother continues to nurture in a limited way while fulfilling the narrative requirement of disappearing in order for Molly to have her own adventures. She grows her branches into a platform for Molly to sleep on, and she produces fruit for Molly to eat that are a curious amalgamation of Molly's favourite flavours and the food that she, as Molly's mother, deems appropriate: "Of course. Mama made it like green beans because she wants me to eat green vegetables" (p. 102). Thus, like Silverstein's giving tree, Molly's mother nurtures Molly through her arboreal form, but unlike the giving tree, she does not sacrifice herself in the process. Molly protects her from their chainsaw-wielding neighbour, and eventually transforms her back.

Although the novel depicts plants as powerful, strange, and significant, capable of "vibrating", "emitting", and "sensing" (Murray, 2015, p. 2), their "desires" are approached only obliquely. Molly's journal documents historical and cultural *uses* for plants and parts of plants, especially herbs—for tea, food, medicine, luck and superstition. She mentions chamomile, apple trees, acorns, mandrakes, borage, lavender, pomegranates, castor oil, and amaranth, among others. The novel explicitly draws on European and pagan botanical traditions, and there is a strong sense that plants have had cultural value for humans for thousands of years. The text is punctuated with small drawings of plants, mostly by Molly. In the novel's closest depiction of a tree's own desires, Pim writes beside his own drawing of a tree: "It takes the weather, and makes food of it. It swallows the sun and rain, drags them inside, turns them into growth" (Murray, 2015, unpaginated). The tree's seemingly magical ability to alchemically transform and consume elements of the natural world again emphasizes the connectedness and kinship of things.

In this novel, then, maternal and arboreal desires converge. Fruit, a common symbol of motherhood and desire, is an expression of Molly's mother's desire to nurture, and Molly's own desire for sustenance and care. What eventually succeeds in transforming Molly's mother back into a human, are Molly's own long repressed tears—physical expressions of her grief and longing for her mother. After her mother has been restored, Molly is glad the adventure has happened for two reasons. Firstly, it is exciting to have been "the child of a tree", and secondly, she is proud to have solved a big problem by herself (Murray, 2015, pp. 222–223). The temporary loss of Molly's mother in the form of a tree prefigures an inevitable permanent loss sometime in the future. The novel thus echoes Jaques's notion of trees as "kindly cipher[s] of nature's paradoxical permanence" (2015, p. 112), as well as Irigaray's experience of the "mothering care of the vegetal world" (2016, p. 21). Relating to her mother in the form of a tree, thus, very literally, being *kin* to a tree, reignites Molly's repressed love for plants. Hugging her mother's trunk, she listens for a "vibration" (Murray, 2015, p. 71). Sleeping outside in the arms of her mother tree, cradled yet alone, Molly forges her own relationship with a "million tiny stars" (p. 154). In the presence of both maternal and arboreal desire and surrendering to the simplicity of her own desire for her mother, Molly opens herself to the strangeness and vibrancy of the world around her. The novel thus affirms a natural connection between mothers and trees, a connection that is problematized in *The Mulberry Tree* and *How to Bee*.

The Mulberry Tree

The Mulberry Tree constructs a far darker vision of maternal and arboreal desire. Ten-year-old Immy and her family relocate to England from Australia while her father recovers from debilitating depression. The family's new house is overshadowed by an oppressive mulberry tree, which, according to legend, steals young girls on the eve of their eleventh birthdays. Two protrusions in its trunk are said to be the bodies of two girls, who disappeared respectively in the 1700s and in 1945. Much of the novel's tension centres around the question of whether or not this legend is true, as Immy awaits her own eleventh birthday with increasing foreboding. While Immy and her family rationally believe the legend is nonsense, the tree itself is described as eerily menacing.

> Halfway up, arm-like branches began to shoot out threateningly, dividing into stout, black fingers which poked and taunted the house cruelly. It was summer, yet the tree displayed no green. Not one leaf. Just inky blackness that blocked out the sky above.
>
> (Rushby, 2018, p. 23)

The tree is more than 500 years old, predating the house and the garden. Drawing on traditions of the "monster plant", from the horror genre (Hall, 2016), it is described as "bewitched", "radiating dark thoughts" (Rushby, 2018, p. 98). It scratches on Immy's window, invades her dreams and appears to hum the haunting tune of an old rhyme. Immy's mother comments that it is amazing that the tree has "lasted as long as it has" (p. 24), revealing an anthropomorphic attitude that assumes humans have the right to kill and dispose of plants whenever the plants are considered useless or unattractive. The estate agent explains that the tree is heritage listed.[2] Its barren, menacing form contrasts sharply with a younger mulberry tree Immy discovers a few streets away. "Instead of being ancient and wizened and cranky, it stood peacefully in its place, leaves fluttering gently in the breeze—tall and green and serene" (p. 124). This tree behaves as a tree should: it remains unthreateningly "in its place", and offers "plump and delicious" berries for all to enjoy (p. 124).

Despite the tree's murderous aspect, Immy displays kindness towards it, and it eventually tells her its story through a series of dreams. It turns out that the mulberry tree's murderous acts occurred out of the anger and bitterness of thwarted maternal desire. The young and fruitful mulberry tree that Immy discovers beside the village green is the "daughter" of the old tree. It had been growing within a few metres of its "mother", but a previous resident had decided to transplant it to the green. This inadvertent act of cruelty can be seen as an act of "plant blindness" (Wandersee & Schussler, 1999)—in assuming that plants are merely the background to animal life, no one considers that the tree may have an attachment to its offspring. In grief and frustration, the old tree retaliates by taking the daughters of the residents of the cottage in place of her own.

The revelation of the monstrous tree as a mother appears to come out of the blue, but actually reflects another source of darkness. Immy's relationship with her depressed father is more central in the text than her relationship with her mother, whose job has brought them to England, but another image of a mother and child haunts the narrative. Immy's parents are doctors, and her father's depression is caused by his guilt about giving an elderly man permission to keep driving, which resulted in a car crash killing a mother and her toddler. Although Immy knows it is sad, she doesn't understand why her father can't move on. The mulberry tree's justifiable supernatural retribution in some way balances Immy's father's unresolvable and unforgivable mistake. While the depiction of the tree's thwarted maternal desire is anthropomorphic, it also draws on Simard's exploration of the role of "mother trees" in forests (2021), and on concepts of plant sentience and personhood. In some ways, the mulberry tree's rage is more appealing than the passive acquiescence to destruction displayed by Silverstein's "giving tree" (1964/2014). A key difference between the two, however, is that while the giving tree's own body is threatened, the mulberry tree's child is removed, and thus anger is perhaps a more socially sanctified maternal response in this instance.

Immy's extension of kinship and kindness towards the tree enables her to understand it, but the tree remains a threat until she brings it a sprig from the branch of its daughter. At this moment, history is changed—the murderous acts of the past are undone. The town is repopulated by descendants of the two girls who now, miraculously, the tree never took, and the tree itself is transformed into a green, spreading canopy, brimming with mulberries to share. The appeased mother tree now feeds, rather than consumes, human children, indicating the possibility of a world where they both can flourish. The possibility of Immy's father's recovery is also implied. Immy's acts of kindness, listening, and "making kin" have satisfied both arboreal, and in this case, vengeful, maternal desire.

How to Bee

How to Bee is set in a dystopian future in which, due to a great famine, there are food shortages, social unrest, unsurmountable class differences, and no wild bees. Peony lives and works on an orchard with her sister and grandfather, and dreams of becoming a "bee". In the absence of bees, a team of children has the job of propagating the fruit trees, by climbing up and down the trees and waving wands dipped in pollen. Peony thus longs to embrace a symbiotic relationship with the trees, fertilizing the flowers to enable the trees to produce fruit. The children who fertilize the flowers are known as "bees", while the children too young, too big, or too clumsy for such a task are called "pests", as they labour to keep caterpillars and insects away from the trees, feeding them to their chickens. Pesticide is no longer used:

> The farm's full of circles. Bees, flowers, fruit. Pests, chooks, eggs. People, bees, flowers, fruit, pests, chooks, eggs, people … all overlapping circles. I don't understand how it went before the famine. Poison? That's like cutting the circles right through the middle. The circle can't go nowhere but a dead end. No wonder the little bees stopped working and left us to starve.
>
> (MacDibble, 2018, p. 14)

In a broken but recovering ecosystem the fruit trees still flourish as the humans both desire and need them: human actions and human names reflect the utter centrality of the fruit tree for human life.

Most of the characters are named after flowers and fruit, indicating the significance of plants in a world in which they have become precarious.[3] In *How to Bee*, the countryside is presented as nurturing and safe, in contrast with the city, which is dangerous and depraved. Peony's mother Rosie works in the city as a maid, and when Peony refuses to join her there, Rosie and her boyfriend kidnap her. The trees in *How to Bee* are presented as places of safety and protection, despite the blunt statement on the opening pages that "sometimes [bees] fall and break their bones" (MacDibble, 2018, p. 2). Peony climbs into a tree to escape her mother: "I wrap my body around it, like I am the tree and the tree is me, and hang on" (pp. 51–52). Peony's mother is not metonymically associated with trees, like Molly's mother, but is rather diametrically opposed to them. For Peony, trees provide the safe, nurturing environment that her mother cannot.

Peony's mother Rosie conforms to the trope of the "unsatisfactory mother" that is common in dystopian fiction (Mitchell, 2016). Her desires are presented by the text as problematic: the desire to live in the city, the desire to earn money, and most damaging of all, the desire to remain in an abusive relationship. Rosie longs for a nuclear family, a house and a steady income. She tells Peony they need to "get ahead" (MacDibble, 2018, p. 78), and extols the benefits of a paid job: "now you'll earn money and we'll be able to rent our own little house and stop living like dogs in the dirt" (p. 126). Peony regards the work they undertake as maids for a rich family as unnecessary and demeaning. She prefers the unpaid but meaningful work she performs in the orchard: "if me and Mags don't do our job the trees suffer, the fruit suffers" (p. 102). The families living and working at the orchard do not earn money, but instead, a place to live and a subsistence enabling them to survive.

The community living at the orchard becomes a life-giving alternative to the defunct model of the nuclear family. Peony escapes the city and returns to the orchard, but her mother stays in the city and ultimately dies in childbirth. This novel, in the inter-generational, interspecies care that it promotes, resonates most closely with Haraway's call to "make kin not babies" (2016, pp. 5–6). Mothers receive short shrift in the narrative. Not only does Peony's inadequate mother Rosie die towards the end of the book, but Lily, the mother of Peony's best friend Applejoy (AJ), who has been sick throughout the narrative, also dies. Peony, AJ, and their younger siblings are left in the care of Peony's grandfather, until AJ's aunt arrives to join the family. The hope at the end of the novel comes in two guises: the safe return of Peony's baby sister after her mother's death, and the revelation that cultivated bees are ready to return to the orchard. Instead of becoming bees, Peony and AJ are invited to become beekeepers. The decidedly non-nuclear, motherless family at the orchard emblematizes care, protection and kinship with the natural world.

Conclusion

While *Molly and Pim* directly overlays the image of the caring mother with the fruitful tree, and *The Mulberry Tree* imputes fierce and dangerous maternal desires to a tree itself, *How to Bee* questions the natural association of the arboreal and the maternal. Here, maternal desire is presented as unreliable, and it is rather the desires of trees that need tending. As Pollan puts it:

> Seeing these plants … as willing partners in an intimate and reciprocal relationship with us means looking at ourselves a little differently, too: as the objects of other species' designs and desires, as one of the newer bees in Darwin's garden.
>
> (2001, p. xxv)

The child protagonists of each of the novels become aware of their identity as "arboreal kindred" (Haraway, 2016, p. 120), and each act in ways to protect and restore the trees they depend on.

The desires of mothers are forces in each of the narratives. Molly's mother's guise as a beautiful and caring tree comes closest to the archetype of the nurturing mother tree, but it is her concurrent desires to mother and to commune with plants that shape Molly's own experience of the world. In contrast, the individualistic desires of Peony's mother are depicted as unsuitable, and she is removed from the narrative. In children's literature, mothers are still required to nurture above all else. *The Mulberry Tree* raises the possibility of surprising and horrifying "arboreal maternal desire". Although it is anthropomorphized beyond recognition, the raw need of the mulberry tree expresses a kind of "appetitive soul" (Plato, 1929). While this does not reveal anything new about what could really be considered "arboreal desire", it does raise the prospect of a tree as a desiring being. In the other two novels, the desires of trees—for earth, sky, bees and rain—are quietly expressed. As the title of *How to Bee* indicates, the novels propose a different way of being in the world, alert to the desires of plants as well as our own.

Notes

1 Interestingly, although all three of these novels are by Australian authors, none of them particularly emphasizes Australian plants. Of the three, *Molly and Pim* refers to the greatest number of specifically Australian plants but concentrates on European and transcontinental plants. *The Mulberry Tree* is set mostly in England, and *How to Bee*, while set in Australia, refers solely to fruit trees of European origin. This observation invites analysis of colonial practices of planting and cultivation, but this falls outside the scope of this chapter.

2 Mulberry trees were first planted in Britain by the Romans. The oldest surviving trees are about 400 years old, and heritage groups have frequently campaigned for their protection from development projects (https://www.moruslondinium.org).

3 This is a theme returned to in MacDibble's next novel, *The Dogrunner* (2019), which imagines a world without grass.

References

Almond, B.(2010). *The monster within: The hidden side of motherhood.* Berkley, Los Angeles and London: University of California Press.

Barrie, J. M. (2015). *Peter Pan: Peter and Wendy and Peter Pan in Kensington Gardens.* London: Penguin. (Original works published 1906 and 1911.)

Bennett, J. (2010). *Vibrant matter: A political ecology of things.* Durham, NC and London: Duke University Press.

Bonaventure, St. (2011). *Mirror of the blessed Virgin Mary.* E-Saint Library. (Original work 13th century.)

Burnett, F. H. (1998). *The secret garden.* New York: Harper Classics. (Original work published 1911.)

de Marneffe, M. (2019). *Maternal desire: On children, love, and the inner life* (2nd ed.). New York: Scribner. (Original work published 2004.)

Fraustino, L. R., & K. Coats (2016). Introduction: Mother's wanted. In L. R. Fraustino and K. Coats (Eds.), *Mothers in children's and young adult literature: From the eighteenth century to postfeminism* (pp. 3–42). Jackson: University of Mississippi Press.

Gaard, G. (2009). Children's environmental literature: From ecocriticism to ecopedagogy. *Neohelicon*, 36, 321–334.

Gibran, K. (2020). On pleasure. *The prophet.* Retrieved from *Poetry Foundation* website: https://www.poetryfoundation.org/poems/148570/on-pleasure (Original work published 1923.)

Grimm, J. L. K., & W. C. (1972). *The complete Grimm's fairy tales* (M. Hunt, Trans.) New York: Pantheon Books. (Original translation published 1944.)

Gutierrez, A. K. (2016). "The hills were in her bones": Living in the blend of mothers and environments. In L. R. Fraustino & K. Coats (Eds.), *Mothers in children's and young adult literature: From the eighteenth century to postfeminism* (pp. 133–152). Jackson: University of Mississippi Press.

Hall, M. (2011). *Plants as persons: A philosophical botany.* Albany: State University of New York Press.

Hall, M. (2016). The sense of the monster plant. In D. Keetley & A. Tenga (Eds.), *Plant horror* (243–255). London: Palgrave Macmillan.

Hall, M. (2019). *The imagination of plants: A book of botanical mythology.* Albany: State University of New York Press.

Haraway, D. (2016). *Staying with the trouble: Making kin in the Chlulucene.* Durham, NC: Duke University Press.

Hines, M. (2004). "He made us very much like the flowers". Human/nature in Anglo–American literature. In S.I. Dobrin & Kidd, K. B. (Eds.) *Wild things: Children's culture and ecocriticism* (pp. 16–30). Detroit: Wayne State University Press.

Irigaray, L., & M. Marder (2016). *Through vegetal being: Two philosophical perspectives.* New York: Columbia University Press.

Jackson-Houlston, C. (2006). "Queen lilies"? The interpenetration of scientific, religious and gender discourses in Victorian representations of plants. *Journal of Victorian Culture* 11(1), 84–110.

Jaques, Z. (2015). *Children's literature and the posthuman: Animal, environment, cyborg.* New York and London: Routledge.

MacDibble, B. (2018). *How to bee.* Warren Barn, West Sussex: Old Barn Books. (Original work published 2017.)

MacDibble, B. (2019). *The dog runner*. Warren Barn, West Sussex: Old Barn Books.
Marder, M. (2014). *The philosopher's plant: An intellectual herbarium*. New York: Columbia Press.
Meeker, D., & A. Szabari (2020). *Radical botany: Plants and speculative fiction*. New York: Fordham University Press.
Merchant, C. (1980). *The death of nature: Women, ecology, and the scientific revolution*. San Francisco: Harper and Row.
Mitchell, J. (2016). "A mom-shaped hole": Psychoanalysis and the dystopian maternal. In L. R. Fraustino & K. Coats (Eds.), *Mothers in children's and young adult literature: From the eighteenth century to postfeminism* (pp. 233–249). Jackson: University of Mississippi Press.
Murray, M. (2015). *Molly and Pim and the millions of stars*. Melbourne: Text Publishing.
Ovid (2004). *Metamorphoses: A new verse translation* (C. Martin, Trans.). New York and London: Norton.
Plato (1929). *Timaeus; Critias: Cleophon: Menexenus; Epistles*. Loeb Classical Library, vol. 234. Cambridge, MA: Harvard University Press.
Plumwood, V. (1993). *Feminism and the mastery of nature*. London: Routledge.
Pollan, M. (2001). *The botany of desire: A plant's eye view of the world*. New York and Toronto: Random House.
Rose, E. C. (1991). The good mother: From Gaia to Gilead. *Frontiers: A Journal of Women Studies*, 12(1), 77–97.
Rowling, J. (1997). *Harry potter and the philosopher's stone*. London: Bloomsbury.
Rushby, A. (2018). *The Mulberry Tree*. Newton: Griffin Press.
Silverstein, S. (2014). *The giving tree*. New York: HarperCollins. (Original work published 1964.)
Simard, S. (2021). *Finding the mother tree: Discovering how the forest is wired for intelligence and healing*. New York: Knopf.
Wandersee, J. H., & Schussler, E. E. (1999). Preventing plant blindness. *The American Biology Teacher*, 61(2), 84–86.
Wardi, A. J. (2016). Maternal ecocriticism and the ecology of motherhood in Jean Toomer's case. *Journal of the Motherhood Initiative for Research and Community Involvement*, 7(1), 22–36.

8 Arboreal Entanglements

Childrenforest and Deforestation in Ecopoetry by Children

Justyna Deszcz-Tryhubczak and Charlotte van Bergen

Childhoodnature is a new conceptualization of childhood and children that has been developing predominantly in educational research in response to the growing understanding of how the anthropogenic impact on the environment is transforming childhood. It constitutes a vital element of nature cultures, a notion proposed by Donna Haraway to name a new way of thinking about the "layers" of human-non-human interrelationalities and interconnectedness (Haraway, 2003, p. 2) that has emerged in such diverse fields as environmental humanities, cultural studies, art, education, childhood studies, and literary studies under the influence of posthumanist and new materialist perspectives. Through questioning Western anthropocentrism, posthumanism and new materialism trouble such entrenched ontoepistemological divides as nature/culture, mind/body, subject/object, reality/language, meaning/matter, and human/non-human, shifting the attention to multiple interdependencies, reciprocities, intensities, energies, and obligations between humans and more-than-human others sharing and creating common worlds. In this view, nature and culture are always already inseparable and entangled through mutually constitutive relations between "the human way of interacting with world" (culture) and "networks of natural systems human biologies are situated in" (nature) (Widdop Quinto & Khatun, 2020, p. 1046). This relational approach also drives the reconceptualization of relationships between childhood and natural environments as childhoodnature: just as adults, children participate in a continuum of mutually shaping and multidirectional connectivities with nature that occur in socioecological contexts involving humans and other kinds of beings and agencies.

Children are nature because they always already belong to more-than-human collectivities. For children, perceiving oneself as a participant in complex ecologies one shares with other entities may mean gaining a sense of agency and understanding of how one's actions and ideas are connected with and influence the environment both here and now and in the future. As childhoodnature can be attended to "through a diverse range of material, social and conceptual practices" (Cutter-Mackenzie-Knowles et al., 2020, p. 10), we propose to see children's literature as part of childhoodnature relationalities that co-shape our awareness of what it is to be in the world. Children's literature, which from a

DOI: 10.4324/9781032066356-12

post-anthropocentric perspective could be viewed as a material expression that "itself produces relationships and whose meaning is produced through human and nonhuman intra-actions" (García-González & Deszcz-Tryhubczak, 2020, p. 51), may become a vital space for children and adults to respond to material conditions of everyday life with care, compassion, and respect for nature. Ultimately, literature is nature, with paper books still being produced with the pulp made of trees.

Childrenforest in a Children's Poetry Assemblage

In this chapter we focus on childrenforest, a concept developed by Anna Vladimirova and Pauliina Rautio to explore an element of childhoodnature manifesting itself in interdependencies of children and forests. In childrenforest, "children's relations to forests [go] beyond the child (in) nature dualism", children's enjoyment of forests or their uniting with them (Vladimirova & Rautio, 2020, p. 335). Childrenforest embodies co-constitutive and spontaneous, if short-termed, entanglements of "the diversity of individuals labelled 'children' and the multiplicity of things and beings called 'forest'" (Vladimirova & Rautio, 2020, p. 337). While studies of representations of childrenforests in texts for young readers focus on literature written by adults (e.g. Jaques 2015; Goga et al. 2018; Guanio-Uluru, 2020), we explore childrenforest through and with a group of 29 ecopoems written by children on human-made deforestation, a salient aspect of the global experience of trees and forests especially in light of the ongoing destruction of the Amazon rainforest and the 2019–2020 bushfires in Australia. Our sample comes from the corpus of ecopoetry written since 2012 by pupils attending John Clare Primary School in the village of Helpston, Northamptonshire, England, the home of the Romantic poet John Clare (1793–1864). Each year all the children write poems on a topic selected by the John Clare Society and participate in a competition it sponsors. Once the theme of the competition is announced, the teachers introduce relevant content to the general curriculum and provide the pupils with mentor texts, including Clare's poems. The children are encouraged to decorate their poems with drawings. The finales of the competition take place during the John Clare Society Festival, organized annually in Helpston. Currently, the circa 600 poems stored in folders at the school's reception are not available to a wider audience.

We look at texts in this collection as not just creative writing exercises, competition entries or children's individual contributions to literary production, but first and foremost as elements within a complex assemblage of human and non-human entities, including the children, the teachers, the parents, ourselves, John Clare and his writings, the children's poems and drawings, the texts that the children read and ideas they exchange, the John Clare Society and the competition, nature around Helpston, and possibly other forces and intensities of which we will never be fully aware.[1] Such an awareness is in fact impossible as assemblages are open-ended and mutate, with agency distributed across them. Mindful of the fact that there are a number of ways in which the Helpston assemblage could be approached, and although we do not know anything about the child authors' actual experiences of forests, we explore one of its elements, childrenforest. Our inquiry is guided both by how the poems on deforestation have been "speaking" with us through unusual, interesting, and unanticipated ideas and images they contain and by our situated position as literary scholars responding to the invitation to reflect on children's literature and plants. Relying on critical plant studies (Kokkola, 2017) and botanical criticism (Ryan, 2018) as elements of our knowledge-making practices, we map the childrenforest co-constituted by the poems by focusing on their arboreal imaginaries; that is, images of tree lives; trees' subjectivity, agency, and affective reactions to deforestation; and the interspecies communities in which trees participate. However, we are not interested in the literary qualities of the poems for their own sake. This research is not "about" the poems, but rather it is conducted "with" the verbal and graphic layers of these texts.[2] Hence, our central research question is how the children's poems invoke childrenforest and how they may attune readers to their own entanglements with trees and forests.

Finally, by focusing on poetry written by children, we also invite a consideration of children's literature itself as an assemblage of matter, discourse, forces, intensities, energies, power relationships, the human and the non-human, whose particular elements, including child authors, can be reterritorialized, for example by centring children's creativity. The posthuman perspective does not negate children's or adults' capacity for self-expression but proposes that "humans are not the only sources of action and that very often their power is sourced from other actors in the vicinity" (Schulte, 2019, p. 99). These other actors may include texts, literary conventions, materials, institutions, and nature. Consequently, child authors' creative agency is one of many expressive agencies within the children's literature assemblage. Such an approach complicates the current research on children's writing as constituting a contribution to children's literature (Conrad, 2016; Cumming, 2017; Wesseling, 2019). It questions in particular the traditional humanist way in which children's production of a literary culture is defined as a component of children's rights. Therefore, although we think of the deforestation poems as standpoints both on the children's own childhoods and on social phenomena in which they participate, we bear in mind that these expressions are part of the Helpston assemblage, which simultaneously generates and is affected by the child authors' personal orientations.

Approaching Assemblages

Researching assemblages carries certain methodological implications. The 29 poems on deforestation were written by the Year 5 and 6 (10–11-year-old) children from Torpel Class for the 2018 competition, whose topic was "The

Wood Is Sweet". We should ideally quote all the poems in full, which is impossible owing to the word limit. While we could analyse two or three poems in detail, we have opted for a flat approach in which we rely on all the poems to develop our argument, quoting relevant passages from them.[3] We have done so for the following ethical reasons. Firstly, by ennobling particular poems as worth special attention, we would be replicating the work of the jury. If we chose different poems than the ones already rewarded, we could be seen as intervening in the competition, which is not our intention. Nor do we want to cause conflicts between the John Clare Society and the school, which has been supportive both of our joint work and of Justyna's broader project related to the collection. Secondly, as these poems are not widely available, a potential reader would be unable to endorse or question our selection by appreciating the other poems from our sample. Thirdly, the children and the parents have given us the permission to study the poems and they might be disappointed that we have centred a particular poem and ignored the others. Finally, we quote the poems the parents have permitted us to disclose, safeguarding the privacy of the children whose consents we did not obtain.

As mentioned above, these poems belong to a larger collectivity, which means, among others, that they resulted from the co-knowing and co-thinking of a particular group of people. Before the children drafted their poems, they reflected on deforestation through the joint exploration of several mentor texts.[4] In *The Vanishing Rainforest* (2007), written by Richard Platt and illustrated by Rupert van Wyk, a child describes how the Yanomami tribe are battling against developers and farmers clearing the rainforest. The book invites the reader to ponder whether it is possible to reconcile the protection of the forest and the preservation of the tribe's lifestyle with the introduction of modern civilization. Lynne Cherry's *The Great Kapok Tree: A Tale of the Amazon Rainforest* (1990) is a picturebook about a man who falls asleep while chopping down a kapok tree. The forest's dwellers—animals and a child—each whisper into his ear about the importance of the tree and the interconnectedness of all living things. When the man awakens and sees all the creatures around him, he abandons his axe and leaves the rainforest. The children also discussed two poems by John Clare—"The Maple Tree" and "The Shepherd's Tree"—and "Trees Please" by Benjamin Zephaniah. Clare—with his characteristic accuracy and physical sensibility—contemplates intimate relationships between the eponymous trees and the phytophylic lyrical-I. Zephaniah's poem lists examples of how trees determine human and animal livelihoods and pleads for leaving them intact. Although Clare's and Zephaniah's poems differ in style and focus, they decentre human agency and stress interdependence of all lifeforms.

The assembled nature of the poems also marks our own experience of them. Justyna became acquainted with the archive when she visited John Clare Primary School in 2018 to collaborate on a different project. Whenever she waited at the school reception, she browsed through the folders containing the poems. Having become intrigued by the diversity of the poems' contents and the creative process in which the pupils engage, she photographed all poems, not knowing yet whether and how she would work on them in the future. Justyna's experience exemplifies "the wonder of data"; that is, "the capacity for wonder that resides and radiates in data, or rather in the entangled relation of data-and-researcher" and is "a counterpart to the exercise of reason through interpretation, classification, and representation" (MacLure, 2013, p. 228). Indeed, the poems kept animating Justyna's thinking after her return to Poland until she applied for funding that would enable her to participate in the creation processes taking place at the school. In the same year Justyna shared the poems with Charlotte, which led to their collaboration on this chapter. Charlotte transcribed all the poems written for the 2018 competition and, together with Justyna, she was reflecting on how to approach them.

All this means that we have not become entangled with specific poems or child poets but with the full assemblage to which they belong. The different elements of the Helpston assemblage have come together to produce thoughts in our minds, making us elements of this collectivity. Analysing selected poems or clearly indicating their authors in our discussion would belie our and their interdependencies within this diverse assemblage. It would also obscure the inter-relatedness of the poems by presenting them as independently existing objects whose meaning is awaiting to be decoded and interpreted. The flat perspective enables us both to convey our experience of the assemblage's distributed agency and to map interconnections among the poems as they enact childrenforest.[5] We hope readers appreciate our attempt to approach the poems in an ethical and methodologically sound way.

Childrenforest in the Helpston Assemblage

Botanification and Anthropomorphism

In all the 29 poems, the lyrical subject is a tree pondering the devastating changes happening around it as a result of deforestation. Presenting trees as contemplating and commenting on the destruction around them, rather than assuming an outsider perspective, demonstrates how the children depart from an anthropocentric viewpoint to use an ecocentric one instead. The reliance on a plant's perspective is referred to as botanification, which "involves a conscious shift away from anthropocentrism and includes moving out of human notions of space and time towards a plant-centric perspective" (McKertich & Shilpa, 2016, p. 46). While the children have probably learnt this technique from Zephaniah's poem, it is used by numerous contemporary poets when they want to "see the world from the perspective of sessile lives" (Ryan, 2018, p. 18). As John Charles Ryan (2018) argues, "vegetal speakers come to life through a material imagination shared by poet and plant" (p. 18). Hence, we propose that, instead of "ventriloquizing vegetal being" (Ryan, 2018, p. 18), the children use botanification to attempt to engage with the plant world and acknowledge the trees as agents feeling possible feelings and thinking possible thoughts.

Moreover, depicting the trees as telling their side of the story and viewing deforestation through their eyes are also ways of generating empathy. Having these arboreal creatures speak for themselves, the poems capture the terror which the trees are experiencing powerfully and convincingly. The empathy for the trees is enhanced by the reoccurring theme of personal loss. The children show the trees as members of larger families, having close bonds with the neighbouring trees. Moreover, the trees addressing the reader are usually the last surviving members of these communities. One tree laments that "as my surroundings disappear it's no longer my home. It was our place but now I'm all alone". Another one cries: "I see my animal friends all run away, they run and run, 'Take me with you'! I say". Still another tree mournfully explains: "As a life can come and go I have more of a scare, my life, my family, my friends are [no] longer there". Some trees pray for their loved ones to be spared: "I pray to god, me, my friends and family will not die tonight and be destroyed, that is just not right".

By envisioning the trees as having mothers, fathers and siblings and as experiencing the human fear of losing them, the poems resort to anthropomorphism. As a stylistic device, anthropomorphism has been criticized for insinuating that the human experience is the only experience. We could, as poet John Shoptaw (2016) explains, indeed sceptically ask when imagining an animal or a plant to be fearful or mournful, "how … we know that their fear is the same as our fear"? (p. 404). According to Shoptaw, however, there is no difference between envisaging a fellow human to feel a certain way and depicting a non-human creature as experiencing certain emotions. The act of imagining something or someone reacting in a specific way is always an act of empathy even, we might add, if it may be more challenging to empathize with non-human life forms than with humans. Rather than projecting the human experience onto non-humans, anthropomorphism offers a way "to empathize beyond humankind" (Shoptaw, 2016, p. 404), showing that the human experience is not unique, but that it could be shared by others, creating interspecies connections and taking the human away from the centre of all worldly being. Through the act of anthropomorphism, the children wonder what losing so many of your own kind feels like: some trees anxiously despair about the inevitable loss of their family and friends; others lament the losses that have already occurred. One tree misses the mutual care it felt as the branches of the other trees touched it lovingly; another one remembers "holding hands with others, sisters, brothers, fathers, mothers". One tree also asks: "Burnt to the ground, to a crisp, I see. Who could do this to my family"? Since the children imagine the trees as experiencing fears that might be close to their own, they narrow the divide between culture and nature and human and tree.

Species Relations and (inter)Dependencies

The arboreal lyrical speakers are also envisaged as members of broader communities.[6] They are relational beings continually interacting with other creatures, be it plants, animals or humans. They provide a shelter, protection, shade, food, and of course oxygen ("I give oxygen, I am a treasure"; "For all the oxygen we have given you for your survival ...") for all the creatures that roam the earth, which the poems often highlight. The trees express their sense of being especially important to animals: "animals begin to climb up me very high, they make a home as I am safe and dry"; "animals crawling up my back, I'm a home to all these creatures, big ones, small ones, lots of screeches". They send a warning to humans that "if you take me, you take the animals too, there will be nothing left except you". The trees also explain their importance to human beings. One tree indicates that "as I stretch my arms out wide, I make shade where you can hide"; another tree invites the reader to engage in a bit of reflection: "Just think a minute what I make, all that fruit that you take. Bananas, pineapples and lots more, all that stuff that you adore".

The poems position trees as important beings that do not merely form the background of human and animal life but actively engage in it. Yet, this dependency does not work both ways ("you take what you want but you give nothing back"), and the extent to which humans depend on trees weighs heavily on these arboreal creatures, eventually resulting in extortion:

> One by one, my family and friends disappear before my eyes,
> Chopping us down for money, medicines and supplies.
> If you cut our civilisation down, you will be effecting [*sic*] yourself also,
> But you still take more, your greed will still grow.

While plants, animals and humans rely on trees for numerous reasons, the latter appear to be autonomous creatures that do need us as much as we need them. By portraying trees as independent creatures, providing for others but being in no need of aid themselves, the poems question the hierarchy of organisms traditionally placing rocks and stones at the bottom and humans at the top. People, who depend on other species and materials for survival, seem infinitely inferior to trees. In the light of critical plant studies and botanical criticism, which diagnose "plant blindness" (Ryan, 2018, p. 6) as characteristic of how culture reflects the crucial role of plants in determining the life on the planet, the child authors do not see trees as lower, passive, and non-sentient forms of life but appreciate trees' "being-in-the-world" (Ryan, 2018, p. 4) and create detailed, often intimate, portrayals of the agency of trees as co-shaping the existence of other beings.

Although the poems depict multiple types of relations between trees and other organisms, they centre the relationship between people and trees, since after all it is human readers that the trees are addressing. While the poems challenge human dominance, they nevertheless portray humans as the "masters ... of the earth's destiny on behalf of human/non-human species" (Cutter-Mackenzie-Knowles et al., 2020, p. 48). Seeing themselves as defenceless, the trees "beg" the loggers to leave them alone. One tree realizes that there is nothing it can do to prevent deforestation and that it is at the mercy of humankind:

> I come from the rainforest so damp and deep,
> I have no protectors, no defence, sadly I weep.
> This is unbalanced, unfair, I can't do anything,
> I am worried by the man, their axes they bring.
> If I could do anything I would, but I cannot,
> They walk towards me, I fear I will rot,
> I am scared and unable to defend,
> When will this terrible torture end?

In another poem, a tree begs humans for mercy:

> Save me, save me!
> Hear my plea!
> I beg you humans,
> Have mercy on me!
> No more, no more!
> No more axes and saws!
> I beg you, I beg you, leave me be!

Deforestation and Environmentalism

The children also reflect on human–tree relations by imagining the trees' direct interactions with humans and describing the latter as "rampaging" and "barbaric" "wood hoggers", who "take pride in swinging that axe", turn "once what was joy into misery", and take a tree's "soul, destroying [its] life their ultimate goal". The destruction of trees by humans creates an overwhelmingly apocalyptic atmosphere that pervades all the poems. It is intensified by the immediacy of the depictions of the trees facing their imminent death. One poem opens with the insistent sounds of chopping—"Chop, chop, chop the wood cutter's here"—with the urgency of the threat being highlighted through repetitions: "Every second of every minute, every minute of every hour, our lives are at risk in the wood chopper's power". Another poem dramatizes a tree's death: "This is my last moment to treasure, I wish it could last forever. Now it is time to say goodbye, I wish I could see this last night ..." In yet another poem, the tree is being burned in the "the red flower" of fire: "The treacherous flower, I ultimately receive; curling through my branches; yet I don't know what's happening to me". Facing their inevitable death and stating their worries about the fate of the planet, the trees expose the damage they endure, turning the poems into a call for justice against people's wrongdoings. Readers may find it disheartening that no poem follows the uplifting scenario from *The Great Kapok Tree*. Contrary to the human in Cherry's story, the people depicted by the young poets are beyond redemption, which might result from the 20 years' difference between the publication of Cherry's story and the poems. As today's children are witnessing adults' inability to reach a consensus about alleviating climate change, they have reasons to worry about their present and future.

The poems' preoccupation with "the changing relationship between human societies and natural environments" (Lidström & Garrard, 2014, p. 37) through an ecocentric lens reveals the children's attitude towards forests and deforestation to be environmental. However, it is also actively environmentalist. A tree recounts looking "over at my fellow trees as they disappear" and another one describes "humans … carrying a deadly axe, chopping all of us down in their path", comparing the act of deforestation to persecution or genocide: "they just kill every single tree. I save all the animals but the logger chops me—I am the great, gone kapok tree". Regarding trees as "earth's precious property", another tree refers to deforestation as theft: "Axes you bring and wood you take, trees you destroy and hearts you break. I used to feel the water on my leaves, but you've taken that away, you ghastly thieves". Asking readers whether deforestation is just, one tree states that this kind of damage and destruction should not be permitted as it means destroying and essentially killing the planet. Depicting "the environment as affected by human activity in an urgent and unsettling way" (Shoptaw, 2016, p. 401) and calling for environmental justice, the children's poems aim "at stimulating the development of a new environmental imagination in its readers" (Shoptaw, 2016, p. 401), which, Shoptaw claims, is key to environmentalist poetry.

By portraying humankind as dangerous and uncivilized, the poems also contest the general image of forests in Western thought. The trees as envisioned by the children—"grand", "great", "glorious", "wise", "tall", "old", and "humble"—seem a lot more cultured than the people in their poems. This approach to the trees might have been inspired by Clare's admiration for their grandeur and beauty, expressed in the poems mentioned above. For a long time, however, forests have represented that which stands outside of culture and civilized society, "the territory—material and imagined; concrete, symbolic and metaphysical—that lies outside the borders of the social contract, the space of the civic and the realms of reason", symbolized by the city (Tavares, 2018, p. 162). The undomesticated, stateless and lawless wild zone of the forest was controlled by the burning and cutting of trees, aimed at providing room for urbanization. The children's poems turn this thinking around. It is not trees but humans that pose a threat, causing destruction wherever they go. Instead of scary, forests are imagined by the children as a safe shelter for all creatures. Envisioning trees as cultured and humans as uncivilized, the poems destabilize the nature/culture binary, acknowledging an entanglement between nature and culture.

Celebrating and Problematizing Nature

While the children's poems have an environmentalist feel to them, they also paint a celebratory picture of nature through the memories the trees wish to share about the past times when forests were still intact. One tree remembers how "there used to be butterflies fluttering around me", whereas others recall how the rainforest, "this once beautiful land", "is now all ashes, though it used to be green", or how they "used to hear all of the rainforest's creatures, the reptiles, mammals and their distinctive features". Reminiscing about better days, the trees "think back to those times, a forest full of greenery". One tree says "farewell" to "the smell of succulent fruits, replaced with smoke from [its] leaves to [its] roots". This celebration of a past nature, as opposed to humankind's current dealings with it, is elaborated by Goga et al. (2018) through the *Nature in Culture Matrix* (pp. 12–14). The matrix enables a text to be located alongside two axes, as propagating either an anthropocentric or ecocentric view of nature and as either celebrating or problematizing nature. Celebration of nature is also connected to ecophenomenological poetry (Lidström & Garrard, 2014), which focuses on nature as a beautiful and fascinating backdrop of human life. However, in the children's poetry, it is not humans that enjoy these natural surroundings. The trees themselves admire the beautiful forest that is their home, thereby creating a non-anthropocentric image celebrating nature. The kind of nature the trees are praising is one that has been undisturbed by human activity—a "perfect paradise … still out of reach, where there's no more fearing, no more screech"—which corresponds to the children's problematization of nature as affected or destroyed by humans. The poems instil the idea that trees would fare better without humans, which is made clear by the trees' constant asking humans to leave them alone ("us rainforest trees, we were perfectly fine, until you came, it was a sign") and their conviction of people's inability to feel "true love for nature and trees". Suggesting that nature is better off left alone, the poems check their own enthusiasm about multispecies connectivities, thereby offering a nuanced view of nature cultures when care and respect are difficult to achieve.

Significantly, in their nostalgia for happier days, the trees recollect a "past that has been sanitized/romanticized" (Cutter-Mackenzie-Knowles et al., 2020, p. 11), turning their joyful memories into "fall-recovery narratives" (Dickinson, 2013, p. 315) to emphasize the destructive attitude that humans, whether deliberately or unintentionally, harbour towards nature nowadays. The idea that relations between humans and nature used to be wholesome would in fact disguise "a long history of environmental degradation and disconnectedness" (Dickinson, 2013, p. 321). However, through the trees sharing their positive memories, the children highlight the severity of the current situation, calling for changing our treatment of nature. By doing so, they honour one particular tree's request: "My hope, my wish, my future dream, is not for trees to shout and scream".

Drawing Trees

While the children's poems express a sense of irrevocable loss and the premonition of a destruction of all life on earth, very few drawings convey equally negative emotions and ideas. Whereas some poems contain images of deforestation, such as smoky fires, burnt trees, trees that have been chopped down, or human figures wielding an axe or an electrical saw, most drawings offer a happier picture of nature than the texts imply. These celebratory images include tall trees with bright green crowns, flocks of birds (classically drawn as little v's), green or autumn-coloured leaves, luscious green grass, flowers in multiple colours, and an occasional butterfly. The way in which the children draw nature may reflect the fact that certain representations, such as the pastoral or the idyll, have "decisively shaped our constructions of nature" (Nyrnes, 2018, p. 75), also becoming elements of childrenforest and childhoodnature.

The drawing in Figure 8.1 exemplifies the rare alliance between the visual and the verbal in the children's poems. The drawing in Figure 8.2 resembles many of the other drawings. Their overall similarity could indicate that the children, and perhaps people in general, have been taught to draw the natural environment in a certain way, following a template that celebrates rather than problematizes nature. This may be connected with the children's focusing on trees in general as only one poem mentions a particular species (the kapok tree). Could this signal the children's inability to recognize tree

Figure 8.1 "An Empty Space in the Greenery". Rendered with permission.

Figure 8.2 "Help Me!" Rendered with permission.

species and substantially limit the poems' ecocentric power? Or do the children resist "reductive scientific terminologies" (Ryan, 2020, p. 100) that instrumentalize human-flora relations? As we can only speculate why most of the young authors did not underscore their poetic ideas in a graphic form, we would like to think that the drawings of healthy trees, a kind of *locus amoenus*, indicate both a lost past and a future that may still happen if humanity reduces its negative impact on the environment.

Conclusion

Childrenforest is "a living, throbbing, productive, open whole rather than [one] composed of interacting (turn-taking), clearly discernible individuals" (Vladimirova & Rautio, 2020, p. 337). In this chapter, we have looked at complex and contradictory, and yet interrelated, ways in which children–forest relationalities emerge from the arboreal imaginaries in the poems about deforestation created at John Clare Primary School. These imaginaries rely on botanification and anthropomorphism emphasizing trees' agencies and affectivities to both confirm and challenge human domination over forests. In this way they provide an account of trees' experiential worlds and modes of being. They also reveal the intricate entanglements binding the fate of trees to that of humans and non-humans, provoking a reflection on what it means to share the world with other species in times of contingency and uncertainty. We hope that by drawing attention to the Helpston assemblage, itself participating in childhoodnature and natureculture, as well as by relating it to the assemblage of children's literature, we have encouraged further considerations of children's and adults' botanical interdependencies as constituted not only by texts addressed to children but also by those written by them.

Acknowledgement: We thank John Clare Primary School in Helpston for enabling us to access the poems. We also thank the child authors and their parents and guardians for their consent to cite the poems. Justyna's contribution was possible thanks to the funding from the National Agency of Academic Exchange (Poland), which she received for her project "Ecopoetic Entanglements: Children's Poetry Mobilizing John Clare's Artistic Legacy". Finally, we thank David Whitley for his feedback on this chapter.

Notes

1 For example, pupils' strategies for writing a poem likely to succeed in the competition. We thank Joe Sutliff Sanders for this suggestion.
2 We follow Karen Coats's suggestion to leave behind the 1970s debate about children's being either natural poets or apprentice poets. As proposed by Coats, we do not reflect on "what we want children's poetry to be, do, and mean" but wonder "what it actually does" (2013, p. 132).
3 We include two poems as examples of contrasting interactions between the verbal and graphic layers of the poems but we have not selected them for their literary qualities.
4 From Ms Sarah Nicolson's private correspondence with Justyna.
5 Significantly, and perhaps paradoxically in light of our interest in trees, we could say we follow the rhizomatic (horizontal, planar, relational) approach proposed by Gilles Deleuze and Felix Guattari in *A Thousand Plateaus: Capitalism and Schizophrenia* (1987) to counteract what they call an arborescent (hierarchic, vertical, and linear) conception of knowledge production.
6 See Thom van Dooren's *Flight Ways: Life and Loss at the Edge of Extinction* (2014) for discussion of sociality as a common feature of all life.

References

Cherry, L. (1990). *The great kapok tree: A tale of the Amazon rain forest*. San Diego: Harcourt Brace Jovanovich.

Clare, J. (2005). Poems chiefly from manuscript. Project Gutenberg. Retrieved 13 May 2020 from www.gutenberg.org/cache/epub/8672/pg8672-images.html

Coats, K. (2013). The meaning of children's poetry: A cognitive approach. *International Research in Children's Literature*, 6(2), 127–142.

Conrad, R. (2016). "My sole desire is to move someone through poetry, and allow for my voice to be heard": Young poets and children's rights. *The Lion and the Unicorn*, 40(2), 196–214.

Cumming, P. (2017). Introduction to "another children's literature': Writing by children and youth" taking writing by children and youth seriously. *Bookbird: A Journal of International Children's Literature*, 55(2), 4–9.

Cutter-Mackenzie-Knowles, A., Malone, K., & Barratt Hacking, E. (2020). Childhoodnature: An assemblage adventure. In A. Cutter-Mackenzie-Knowles, K. Malone & E. Barratt Hacking (Eds), *Research handbook on childhoodnature: Assemblages of childhood and nature research* (pp. 1–15). Cham: Springer.

Deleuze, G., & Guattari, F. (1987). *A thousand plateaus: Capitalism and schizophrenia*. Minneapolis: University of Minnesota Press.

Dickinson, E. (2013). The misdiagnosis: Rethinking "nature-deficit disorder". *Environmental Communication: A Journal of Nature and Culture*, 7(3), 315–335.

García-González, M., & Deszcz-Tryhubczak J. (2020). New materialist openings to children's literature studies. *International Research in Children's Literature*, 13(1), 45–60.

Goga, N., Guanio-Uluru, L., Oddrun Hallås, B., & Nyrnes, A. (2018). Introduction. In N. Goga, L. Guanio-Uluru, B. Oddrun Hallås & A. Nyrnes (Eds.), *Ecocritical perspectives on children's texts and cultures: Nordic dialogues*. Basingstoke: Palgrave Macmillan.

Guanio-Uluru, L. (2020). Imagining climate change: The representation of plants in three Nordic climate fictions for Young Adults. *Children's Literature in Education*, 51, 411–429. https://doi.org/10.1007/s10583-019-09387-4

Haraway, D. J. (2003). *The companion species manifesto: Dogs, people, and significant otherness*. Chicago: Prickly Paradigm Press.

Jaques, Z. (2015). *Posthumanism and children's literature: Animal, environment, cyborg*. New York and London: Routledge.

Kokkola, L. (2017). Critical plant studies in children's literature. In C. Beauvais & M. Nikolajeva (Eds.), *Edinburgh companion to children's literature* (pp. 274–280). Edinburgh: Edinburgh University Press.

Lidström, S., & Garrard, G. (2014). "Images adequate to our predicament": Ecology, environment and ecopoetics. *Environmental Humanities*, 5, 35–53.

MacLure, M. (2013). The wonder of data: Cultural studies. *Critical Methodologies*, 13(4), 228–232. https://doi.org/10.1177/1532708613487863

McKertich, P. V., & Shilpa, V. (2016). Plant poetry and the botanification of the imagination. *Journal of Literature and Science*, 9(2), 36–49.

Nyrnes, A. (2018). The Nordic winter pastoral: A heritage of romanticism. In N. Goga, L. Guanio-Uluru, B. Oddrun Hallås & A. Nyrnes (Eds.), *Ecocritical perspectives on children's texts and cultures: Nordic dialogues* (pp. 75–89). Basingstoke: Palgrave Macmillan.

Platt, R., & Van Wyk, R. (2007). *The vanishing forest*. London: Frances Lincoln Children's Books.

Ryan, J. C. (2018). *Plants in contemporary poetry: Ecocriticism and the botanical imagination*. New York & London: Routledge.

Ryan, J. C. (2020). Writing the lives of plants: Phytography and the botanical imagination. *a/b: Auto/Biography Studies*, 35(1), 97–122. https://doi.or/10.1080/08989575.2020.1720181

Schulte M. C. (2019). Wild encounters: A more-than-human approach to children's drawing. *Studies in Art Education*, 60(2), 92–102. https://doi.org/10.1080/00393541.2019.1600223

Shoptaw, J. (2016). Why Ecopoetry? *Poetry*, 207(4), 395–408.

Tavares, P. (2018). Forest. In R. Braidotti & M. Hlavajova (Eds.), *Posthuman glossary* (pp. 162–167). London: Bloomsbury.

Van Dooren, T. (2014). *Flight ways: Life and loss at the edge of extinction*. New York: Columbia University Press.

Vladimirova A., & Rautio P. (2020). Unplanning research with a curious practice methodology: Emergence of childrenforest in the context of Finland. In A. Cutter-Mackenzie-Knowles, K. Malone & E. Barratt Hacking (Eds.), *Research handbook on childhoodnature: Assemblages of childhood and nature research* (pp. 335–360). Cham: Springer. https://doi.org/10.1007/978-3-319-67286-1_25

Wesseling, L. (2019). Researching child authors: Which questions (not to) ask. *Humanities*, 8(2), 87. https://doi.org/10.3390/h8020087

Widdop Quinton, H., & Khatun, F. (2020). Childhoodnature alternatives: Adolescents in India, Nepal, and Bangladesh explore. In A. Cutter-Mackenzie-Knowles, K. Malone & E. Barratt Hacking (Eds.), *Research handbook on childhoodnature: Assemblages of childhood and nature research* (pp. 1043–1074). Cham: Springer. https://doi.org/10.1007/978-3-319-67286-1_57

Zephaniah, B (n.d.). Trees please. The Children's Wood. Retrieved 13 May 2020 from https://www.thechildrenswood.co.uk/benjamin-zephaniah/.

The Children's Poems We Had Permission to Quote:

Harrie Metcalfe, "*The Kapok Tree*".
Hamish Robertson, "*Help Me!*".
Oscar Webb, "*If I Leave*".
Caitlin Walls, "*Scarred*".
Lorri Fleming, "*The Wood Cutter*".
Max Cross, "*Defenceless Trees*".
Sam Brownlow, "*Just Leave*".
Amelia Rogers, "*Soon, No More*".
Emily Fosbeary, "*A Tree's Tale*".
Eva Randall, "*An Empty Space in the Greenery*".
Lily Hutton-Smith, "*Ashes in the Air*".
Emma Dennis, "*Let Us Be*".
Isobel Mason, "*Don't Cut Me Down*".
Jessica Brownlow, "*A Prayer of a Tree*".
Paige Grant, "*Goodnight My Trees*".
Isla Baker, "*Farewell*".
George Abell, "*Last Breath*".
Lucy Munday, "*The Fallen*".

9 "I felt like a tree lost in a storm"—The Process of Entangled Knowing, Becoming, and Doing in Beatrice Alemagna's Picturebook *Un grande giorno di niente* (2016)

Nina Goga

Although critical plant studies is a fairly new discipline within the humanities, trees and forests have been a long–standing, crucial, and significant component of humans' existence in the world, be it as a resource for nutrition, a source of heat or tools, as shelter from enemies or harsh weather conditions, or as a companion one may seek comfort in, or with whom one may share existential, religious, or philosophical worries, doubts, and dilemmas. Consequently, there is a vast and long oral, as well as written, folkloristic, as well as scientific, tradition of expressing humans' ideas about trees and forests.[1]

In this chapter I will examine how the child character in the picturebook *Un grande giorno di niente* (2016a, *On a magical do-nothing day*) intertwines with trees and other nature materials.[2] The book was created by the academically unexplored Italian–French artist Beatrice Alemagna, and may be perceived as a boy's report on one specific day when he, after playing for a long time at his game, is urged by his mother to go outdoors in the rain.[3] Although he brings along his game, he must soon manage without it when he loses it in a pond. The rest of the day he roams the woods experiencing new qualities of both himself and the various life forms that he encounters and engages with. Safely home again, he seems to re-connect with his mother and himself as a child inevitably interwoven with the neighbouring nature.

Alemagna's picturebook was selected for this investigation of a tree–child relationship for at least two reasons. The first is because of the artist's original blending of "a rare depth of visual literacy with gentle, poetic humanity and a fearless approach to experimenting with media and materials" (Salisbury & Styles, 2012, p. 66).[4] Secondly, but not less importantly, I have chosen the picturebook because of the plot and the particular double-spread which initiated my curiosity, that is, the doublespread where the I narrator declares that he "felt like a tree lost in the storm" (Alemagna, 2016a, doublespread 8, see Figure 9.1).[5] Not only the connective act reflected in the narrator's thought, but also the way this entanglement is represented in the illustration absorbed me.

DOI: 10.4324/9781032066356-13

Figure 9.1 *Un grande giorno di niente* (2016) by Beatrice Alemagna, Milano: Topipittori. Reprinted with the artist's permission.

Drawing on theoretical perspectives on the tree–human relationship in children's literature, posthuman child-oriented theory, and ecocritical studies of children's literature, the aim of the chapter is to map out how various forms of nature/culture binaries are depicted and blurred in the book and how the boy intra-acts together with trees and other life forms, in both the verbal and the visual text. An overall hypothesis is that through these various intra-actions the boy restores his conception of outdoor life, and of his relationship or kinship with it.

In line with current studies concerning posthumanism and the child, such as *The Posthuman Child: Educational Transformation through Philosophy with Picturebooks* (Murris, 2016) and *Posthumanism and Literacy Education* (Kuby, Spector & Thiel, 2019), the analysis of Alemagna's picturebook will be conducted as a *diffractive reading*, that is, a reading aiming to shift the focus away from questions of human knowing "toward questions of entangled knowing/becoming/doing in relation with other lively matters and meanings" (Kuby et al., 2019, 10). Hopefully, this way of reading will lead to an increased awareness of how children imagining themselves as plants in children's literature destabilize the traditional tree–human relationship where the human/the child is placed "in a position of mastery, as husband to the forest" (Jaques, 2015, p. 118).

The Tree–Human Relationship in Children's Literature

Child characters in children's literature connect with woods and trees in many ways and in previous studies woods and trees have often been treated together without much clarification of the symbolic, mythological, or ontological nuances between them. More often studies tend to start with discussions of woods and move to occasional readings of eventually significant trees (Borg & Ullström, 2017; Winters, 2018; Christensen, 2019; Jönsson, 2019). In a recent study, I (Goga, 2020) identified some characteristic ways of representing the interplay between child characters and woods and trees, that is, woods and trees as a place of excursion, of refuge or freedom, as a homeplace, and as a place of knowledge and of reflection. I also noted that the child characters were depicted in rather conventional ways both verbally and visually, in accordance with the various characteristics. For instance, when woods or trees were depicted as a place of knowledge and/or reflection, the child character was often sitting on a branch leaning towards the trunk with her/his legs curled up. Or if woods and/or trees were linked to freedom, and perhaps also to some sort of rebellion, the child was hanging upside down from a branch.

According to Zoe Jaques (2015), who is one of the children's literature researchers who has investigated tree–human relationships the most thoroughly,

> trees have long held a kind of nascent posthuman potential. They point to the ways that environment is of superior importance to the individual life of man, while also, as a source of heat and shelter, sustaining that very life. Trees can, therefore, be said to have the most symbiotic of relationships with humans, and they can also generate as much affection as pets, thereby deconstructing a humanistic hierarchy of being which traditionally places fauna above flora.
>
> (Jaques, 2015, p. 115)

While Jaques has studied how sexual danger, attraction, and devotion are inscribed in myth-inspired arboreal metamorphosis or girl-to-tree transformations (Jaques, 2013), and how sentient trees[6] are in service to humanity in children's literature (Jaques, 2015), I shall examine a representation of a tree–child entanglement where the tree is not anthropomorphized or in the service of anyone, but lends and blends, without being at risk, its vegetal characteristics with the child.

Diffractive Reading and Agential Cuts

Carrying out a diffractive reading implies paying attention to a text's agential cuts, that is, diffractions where established or unquestioned binaries or hierarchies are destabilized or challenged. The term "diffractive" derives from physics and describes "the way waves combine when they overlap and the apparent bending and spreading of waves that occurs when waves encounter an obstruction" (Barad, 2007, p. 74). Building on Donna Haraway's notion of diffraction, the diffractive methodology has been developed by Karen Barad. While Haraway introduced the notion as a metaphor to help re-think difference/s beyond binary oppositions, Barad insists on the descriptive meaning when stating that diffraction means "to break apart in different directions" (Barad, 2014, p. 168). What is crucial to Barad is that it is where the waves interfere or overlap that they change in themselves and intra-act, hence waves are forever becoming. A diffractive reading, then, might be understood as

> an iterative practice of intra-actively reworking and being reworked by patterns of mattering. A diffractive methodology seeks to work constructively and deconstructively (not destructively) in making new patterns of understanding–becoming.
>
> (Barad, 2014, p. 187, endnote 63)

The concept of diffractive reading has been applied to work with and analyses of children's literature by several scholars within the field of children's literature research and posthuman theory (e.g. Murris, 2016; Kuby, Spector & Thiel, 2019; Burton, 2019). My research builds on the work by Karin Murris (2016) and Candace R. Kuby, Karen Spector and Jaye Johnson Thiel (2019). Murris's main objective is to disrupt the mind/body dualism, which she identifies as the core dichotomy, "and with it the engrained belief that humans' unique ability to reason' place[s] them 'above' and 'give[s] them control' over animals, plants, and their physical environment" (p. 5). She considers picturebooks "as creative opportunities to destabilise discriminatory binaries" (p. 200) such as mind/body but also nature/culture and human/non-human. In line with Barad, Murris explains how intra-action is different from interaction "in that 'nature' and 'culture' are never 'pure', are never unaffected by each other, but are always in relation" (p. 12). Hence, it is impossible to detect or decide where the boundaries are, including, in our case, the boundaries between tree and child.

Other binaries that are often at work in picturebook analyses are those related to word/image and linear/spatial. Instead of ascribing different qualities and affordances to words and images, a diffractive reading seeks to focus on the materiality of the book, e.g. format, paper quality, typography, layout, colours, and artistic style (visual as well as verbal). According to Murris (2016, p. 213), aspects of materiality, such as colours and metaphors, can be performative agents that create or open new cuts that do not privilege, for instance, humans above plants or words above images.

As stated above, a diffractive reading aims to shift the focus away from questions of human knowing "toward questions of entangled knowing/becoming/doing in relation with other lively matters and meanings" (Kuby et al., 2019, 10). This shift of focus should "start with what is *not* familiar" (Murris, 2016, p. 208) or with cuts that are not too small as Kuby et al. (2019, pp. 1–17) phrase it. An example in Alemagna's *Un grande giorno di niente* of something that is not familiar and that may create an agential cut is exposed to the reader in the doublespread where the I narrator, after losing his video game in the pond, leans his back towards a trunk and concludes that "[t]he drops hit me on my back like stones. I felt like a tree lost in the storm" (Alemagna, 2016a, doublespread 8, see Figure 9.1). Both the verbal metaphor and the illustration blur the plant/human or tree/boy binary. The tree and the boy are infused with one another. Their fibres intertwine.

I will compare the unfolding of this doublespread with the moment when waves break apart in different directions. Hence, the doublespreads to follow direct the narrator and the reader to multiple encounters, overlappings and intra-actions with various life forms before they (the boy and the reader) return home and are able to intra-act with his mother and the book as a whole, that is, to re-read it in overlapping and new ways. In my reading, which is also an ongoing re-reading, I will depart from this disruptive doublespread and try to detect the potential binaries hinted at on previous doublespreads and pursued on subsequent doublespreads.

Like a Tree Lost in the Storm

Un grande giorno di niente is characterized by a particular blending of dark, earthy colours (crayons and grease pencils) on thick and rough paper and the strong and signalling colour of the narrator's raincoat (see, e.g. Figure 9.1). This signal orange also covers the endpapers of the book. Another characteristic of the book is Alemagna's extended use of overview, and illustrations where the protagonist is placed in a broader environment, mainly outdoors. Finally, in contrast to so many tree–child representations in children's books where the child character talks to trees, climbs trees, or is embraced and comforted by trees, Alemagna's child protagonist leans into the tree and undergoes, so to speak, a sort of vegetal morphing. Instead of violating, usurping or demanding something from the tree, the boy metaphorically borrows the vegetal qualities of the tree's veins, fibres, bark, and roots (see Figure 9.1).

The above-mentioned justifications for the choice of book and the characterizations of it could be summed up in a recognition of the book as an example of what Lawrence Buell (1995) terms 'environmental literature'. Following his checklist (pp. 7–8), one may claim that the picturebook presents the non-human environment in a way that suggests that human history is implicated in natural history (1), that the human interest is not understood to be the only legitimate interest (2), that human accountability to the environment is part of the text's ethical orientation (3), and that it is implicit in the picturebook that the environment is a process rather than something constant or given (4). Although it is not my aim to follow this recognition in detail, the environmental qualities of the picturebook will accompany my reading.

Before carrying out what I call a diffractive reading of the book, I would like to point out that others may suggest alternative readings, depending on whether they emphasize the child–mother relationship instead of the child–nature or the child–tree relationship. Such readings do not have to contrast with the one I will suggest, they could rather support or broaden it. A reading emphasizing the child–mother relationship will necessarily also have to question the underlying reasons why they seem to have nothing to tell or share with each other at the outset of the story. The first doublespread depicts the two of them when they, on a dark and rainy day, arrive at a large house surrounded by fields and woods. The text reads "We were there for the hundredth time. Me and my mother in the usual holiday house. The usual forest. And the usual rain" (doublespread 1).[7] On the next spread they are depicted indoors in a setting that seems to be the regular one for their stay: the mother at her desk working on her computer, and the boy laying on the sofa killing Martians in his game.

According to the narrator, this is how the days pass. Nevertheless, the living room is somehow cosy, with warm colours, sofa cushions, firewood under a bench, a guitar in the corner, and traces of nature such as small stones, branches of a bush, and a flower in a small vase.[8] In addition to his comments on their doings, the boy reports that he is thinking of his father and of "everything he would have shown me outside. A lot of wonders" (Alemagna, 2016a, doublespread 2).[9] Together with the last two doublespreads of the book concerned with the boy's return home, these two doublespreads frame the story. While the framing story stresses the process of *inter*–action between the child and the mother, the rest of the story is set outdoors and focuses on the process of *intra-act*ion between the child and the many unfamiliar curiosities or forms of life to be re-discovered there. There is no need to keep these stories apart; instead, I think it is enriching to read them as overlapping and entangled in each other. Hence, one may suggest that by engaging with the many life forms or earthlings outdoors, the boy is able to remember and mourn the loss (or absence) of his father and subsequently manage, together with his mother, to overcome the human/nature breaches, fractures or binaries.

I have already mentioned that the second doublespread, depicting mother and child in the living room, hints at the existence of an outdoor nature: the many remnants of trees, plants, and stones and the outdoor wonders that the boy imagines that his father could have shown him. This blending of indoors and outdoors, of culture and nature, is already suggested to the reader before the main story starts. Or perhaps, according to established picturebook theory, the story may start at the very cover of the book (Nikolajeva & Scott, 2006, p. 241). In accordance with the objective of this chapter, I will limit my comments on the picturebook's paratexts to the two doublespreads that follow the front endpaper. The first one displays a full-page illustration of a rusty-red armchair placed under a tree in a meadow. A child, looking at the reader-viewer, lays across the armchair with his head and feet placed on the armrests. He is dressed in a signal orange coat and holds a game in his hands. He is in the company of three snails, one of them on the hood of his coat. At the foot of the chair there are four mushrooms (amanitas) and a dozen stones. In the upper left corner of the doublespread the author has placed her dedications to Francesca "who sees the world shine through the stones", to her "big Mimosa" and "to Tiberio, who inspired this book and to Veronica, Manu, Pata, John B., Claudia, Béatrice and Sandro". Hence, this spread communicates both visually and verbally from the very beginning the environmental ambiguity of the story. Turning to the title page of the picturebook, the reader-viewer encounters another curious blending of nature/culture: both the artist's name and the title of the book are typed in a sort of matrix font, alluding to the game the boy is holding in his hand on the previous page and foreshadowing the Martians that will be mentioned and depicted two pages later. In addition to the artist's name and the title, the page also displays a vignette representing a pair of boots in the grass surrounded by mushrooms in shades of red and orange. Again, the visual and verbal interplay draw the reader-viewer's attention to an entangled natureculture, that is, "a synthesis of nature and culture that recognizes their inseparability in ecological relationships that are both bio-physically and socially formed" (Malone & Ovenden, 2017, p. 1).

Despite the inclusion of nature or various life forms in his indoor environment, the boy seems unsusceptible to engaging with others besides the Martians in his game, which he sneaks with him in the pocket of his rain-coat when he is urged to go outside. In his first attempt to interact with or step into the outdoor environment he his occupied with keeping his game safe and dry. He is depicted in a succession of images in the same double-spread with his face and body turned towards the previous doublespread, like he longs for home. On the next page he is depicted jumping from stone to stone in a pond while comparing the stones to the heads of the Martians. It is only when he loses his game and starts feeling like a tree that he becomes more sensitive to details in nature as they appear in themselves.

By morphing and metamorphizing with the tree, the boy's agency takes new directions and his awareness of the matters around him starts diffracting. The following doublespreads support this coming closer to nature by zooming in on the boy and the snails (doublespread 9) and mushrooms (doublespread 10), and finally by depicting only the boy's hands engaging with the mud: "I felt grains, stones, clumps, roots, and berries swarming under my fingers. An underground world full of unknown micro–matter … which I could touch" (doublespread 11).[10] This intra-action with mud and matter seems to bring the boy to some revelation; sunrays hit him (double-spread 12), he runs and stumbles down a hillside (doublespread 13), and at the foot of the hill he experiences the world as upside down, "[e]verything was as if it were new" (doublespread 14).[11] The materiality of this doublespread is also worthy of some attention. The reader-viewer is invited to investigate this upside down sensation. The boy is laying at the bottom of the doublespread staring at the woods above him. At first one may conventionally think that the boy is laying on the ground looking up at the trees, but by paying attention to the way the trees are depicted one must rethink this conception. One may turn the book upside down and find the boy lifted up in the sky looking down at the trees or one may think of the distorted positions of trees and child as an expression of the boy's impression that the world "could have been re-made from the beginning" (doublespread 14).[12] This upside down positioning may also have something in common with the rebellious child hanging upside down from a branch as is often found in other tree–child relationships.

Up till now both the story and the reading of it have fluctuated between moving forward, doing, and becoming aware of. On the subsequent doublespread the process of knowing seems to spread out. Instead of an overview, Alemagna presents the reader-viewer with an almost nonfictional chart with several nature–child intra-actions and entanglements, most of them including part of a tree: sitting on a branch and looking into the distance, breathing in air while turned towards the trunk, drinking raindrops from pine needles, hanging from a bough and looking at a row of marching insects, and talking with a bird that is sitting on a leafy branch. This manifold *intra*species contact is not the finale, but, in the now of the story, the culmination of the boy's 'process of entangled knowing'. The initial becoming one with the tree and the following doing and intra-acting with various matter seem to have brought him to an ethical state of thinking where humans' accountability to the environment can't be overlooked or hidden away. Thinking, as Murris puts it, "is *part* of 'the' world; an intra-active engagement with specific configurations of the world" (Murris, 2016, p. 17). I will add that this state of thinking shares the characteristics of a posthuman ethics, that is, an ethics which "is mutually entangled with place, culture, multispecies, and matter. As well as all the social and political implications that are born out of the liveliness and energy produced through these entanglements" (Kuby et al., 2019, p. 13). Consequently, it is only by experiencing the world as new and re-constructed from the beginning, by entangling with the trees' fibres and by disrupting the binaries between nature/culture and inside/outside that the boy manages to disrupt the distancing silence between him and his mother and establish a tacit, but appreciating connection between them when they, in the final doublespread, are able to look into each other's eyes and faces while drinking hot chocolate at the kitchen table in a kitchen where warm red colours are scattered across the page.

Concluding Remarks

In this chapter I have studied how a specific tree–child relationship, expressed metaphorically by the narrator of Beatrice Alemagna's picturebook *Un grande giorno di niente*, disperses its metaphorical agency and influence and infuses the reading of both previous and subsequent parts of the book. By carrying out a diffractive reading aimed at focusing on questions of entangled knowing/becoming/doing in relation with other lively matters and meanings, I found that the metaphorical and visual fibrous entanglement of boy and tree redirected the boy to engage with his surroundings in unfamiliar and unexpected ways, which in turn opened up an ecological engagement with all other life forms, including his engagement with his mother and the domestic environment they share. Following the woods-roaming boy on his environmental awareness adventure, the reader-viewer is invited to partake and engage in the becoming/doing/knowing and to re-think or re-do his/her own natureculture practices.

Notes

1 For an introductory overview of mythical and religious texts and thoughts dealing with trees, see for instance Manuel Lima and Ben Shneiderman's *Book of Trees. Visualizing Branches of Knowledge* (2014), and for scientific and historical perspectives on trees, see for instance Peter Thomas's *Trees: Their Natural History* (2000).

2 A few words about translation: The book was first published in French in 2016 and shortly after, also in 2016, in Italian. Since the French and the Italian versions of the text were written by the artist, I will consider both versions as original picturebooks. There are some critical differences between the French and Italian versions which I consider to be an expression of the artist's amendment of the picturebook. Although the book has been translated into English from the French version, I will refer to the Italian edition, which I consider the most complete version due to a more appropriate us of for instance "the drama of the turning page" (see endnote 5 for an example). Additionally, the English translation deviates and differs from the French in many passages. All translations of the text from Italian into English are mine. When I quote from the text further on in this chapter, I will render the text both in Italian and in French and, when relevant, I will comment on the differences between the various versions, including the English translation. Although it is worth a study of its own of how something like environmental awareness gets lost in translation, this chapter does not intend to be a complete translation study. Many of the differences between the English translation and the French or Italian originals seem to be affected by the prevailing ideologies and values in the target culture. As indicated by Riitta Oittinen, Anna Ketola, and Melissa Garavini (2018) in their study on picturebook translations, the translators are always affected by how they understand the needs of the receiver of the translation.

3 I refer to the I narrator as a boy due to a few adjectives and verbal tenses in the text that indicate that the character is of the male gender. For example, on doublespread three where the narrator tells that he went out, "sono uscito", the 'o' in 'uscito' indicates that the narrator is male. In the English translation or in translations to, for instance, Scandinavian languages the gender of the character would not be possible to trace linguistically. Although I do not comment any further on the gender aspect, it may be worth analytical attention. As pointed out by one of the peer reviewers the boy's ways of acting outdoors may be a way to renegotiate hegemonic masculinity.

4 Alemagna's picturebooks have won several awards. The selected picturebook has received the Huckepack Prize (Germany) in 2019, the English Association Book Award in 2018, the Gold Medal of The Original Art Exhibition of the Society of Illustrators (USA) in 2017, the Grand Prix de l'Illustration (France) in 2017, and the Landerneau Prize (France) in 2017.

5 "Le gocce mi picchiavano sulle schiena come sassi. Mi sentivo un albero perso nella tempesta" (Alemagna, 2016a). In the Italian version this is the complete text on this doublespread. The full text on this doublespread is somewhat different in the French version and the English translation, which read respectively, "Et maintenant, qu'allais–je–faire sans mon jeu? Les gouttes cognaient comme des pierres sur mon dos. J'étais un arbre perdu dans la tempête" (Alemagna, 2016b) and "Without my game, I had nothing to do. The rain was so hard it felt like rocks hitting me. I was a small tree caught in a hurricane" (Alemagna, 2017). Instead of stating that he *is* a small tree, as in the French and English texts, the narrator in the Italian version describes a feeling: "I felt like a tree lost in the storm". The confirmative use of 'I was' makes the experience static and misses the process of becoming. In addition, the first sentence in the French text appears in the Italian version on the previous doublespread, which seems more in line with established picturebook theory and the idea of "the drama of the turning page".

6 Examples of sentient trees in Jaques's study are found in *The Song of Hiawatha* (1885), *The Giving Tree* (1964), and in the Harry Potter series (1997–2007).

7 "Eravamo lì per la centesima volta. Io e mia madre nella solita casa di vacanza. Con la solita foresta. E la solita pioggia" (Alemagna, 2016a) / "Nous y étions. Pour la deuxième fois. Ma mère et moi dans la même maison de vacances. La même forêt. Et la même pluie" (Alemagna, 2016b). To this passage the English translation has added the following sentence: "Dad back in the city" (Alemagna, 2017). This is critical to the interpretation of the book, since it is never spelt out why the father is not there. By claiming that the father is back in town, the English translation limits the reader's interpretative space. There are no safe traces in the original versions confirming whether the father is only absent, like being back in the city, or separated from the boy, for instance because the parents are divorced, or finally, whether the father is dead.

8 According to Marnie Campagnaro (2019), Alemagna's domestic geographies are often emotional maps of parent–child relationships.

9 "a tutto quello che mi avrebbe mostrato fuori di qui. Un sacco di meraviglie" (Alemagna, 2016a) / "à tout ce qu'il m'aurait montré dehors, avec son sourire émerveillé" (Alemagna, 2016b). Once again, the English translation differs from the French and Italian versions: the whole passage about him thinking of his father is left out. Instead, the following sentence is added: "I wished Dad were here" (Alemagna, 2017).

10 "ho sentito grani, granelli, grumi, radici e bacche brulicare sotto le dita. Un mondo sotterraneo pieno di micro–cose sconosciute... potevo toccarlo!" / "graines et granules, noyaux, grumeaux, racines et baies fourmillaient sous med mains. Un monde souterrain plein de microchoses inconnues [...] Je pouvais le toucher!"

11 "Ogni cosa era come nuova" / "Chaque chose me semblait inconnue".

12 "Sembrava che tutto fosse stato rifatto da capo" / "On aurait dit que tout avait été refait à neuf".

References

Alemagna, B. (2016a). *Un grande giorno di niente.* Milan: Topipittori.

Alemagna, B. (2016b). *Un grand jour de rien.* Paris: Albin Michelle Jeunesse.

Alemagna, B. (2017). *On a magical do-nothing day*. London: Thames & Hudson.
Barad, K. (2007). *Meeting the universe halfway: Quantum physics and the entanglement of matter and meaning*. Durham, NC: Duke University Press.
Barad, K. (2014). Diffracting diffraction: Cutting together–apart. *Parallax*, 20(3), 168–187, doi:10.1080/13534645.2014.927623
Borg, C. B., & Ullström, M. (2017). The child in the forest. Performing the child in 20th century Swedish picture books. *LIR.journal*, 9, 7–24. Downloaded from http://ojs.ub.gu.se/ojs/index.php/LIRJ/article/view/4126/3377
Buell, L. (1995). *The environmental imagination: Thoreau, nature writing, and the formation of American culture*. Cambridge, MA: Harvard University Press.
Burton, L. (2019). The posthumanist child: Pharmakon and Collodi's Pinocchio. *Oxford Literary Review*, 41(2), 202–218.
Campagnaro, M. (2019). Geografie domestiche e visioni di "genere" negli albi illustrati. In G. Zago, C. Callegari & M. Campagnaro (Eds.), *La casa. Figure, modelli e visioni nella letteratura per l'infanzia dal Novecento ad oggi* (pp. 79–108). Lecce: Pensa Multimedia.
Christensen, S. (2019). Without a word: Bonding with the northern nature in Lene Asks's *Du*. *Nordic Journal of ChildLit Aesthetics*. doi:10.18261/issn.2000-7493-2019-01-10
Goga, N. (2020). Økokritiske perspektiv på representasjoner av skog i skandinavisk barnelitteratur. In U. Nemeth & A. Nordenstam (Eds.), *Svensklärarföreningens Årsskrift 2019* (pp. 113–129). Stockholm: Natur och Kultur.
Jaques, Z. (2013). Arboreal myths: Dryadic transformations, children's literature, and fantastic trees. In I. Gildenhard & A. Zissos (Eds.), *Transformative change in western thought: A history of metamorphosis from Homer to Hollywood* (pp. 163–182). Abingdon: Modern Humanities Research Association and Routledge.
Jaques, Z. (2015). Tree. In *Children's literature and the posthuman: Animal, environment, cyborg* (pp. 111–142). London and New York: Routledge.
Jönsson, M. (2019). Barnen, orden, skogen. *Provins*, 1, 46–50.
Kuby, C. R., Spector, K., & Thiel, J. J. (2019). *Posthumanism and literacy education*. New York: Routledge.
Lima, M., & Shneiderman, B. (2014). *Book of trees: Visualizing branches of knowledge*. New York: Princeton Architectural Press.
Malone, N., & Ovenden, K. (2017). Natureculture. In A. Fuentes (Ed.), *The International Encyclopedia of Primatology*. New York: John Wiley & Sons. doi:10.1002/9781119179313.wbprim0135
Murris, K. (2016). *The posthuman child: Educational transformation through philosophy with picturebooks*. London: Routledge.
Nikolajeva, M., & Scott, C. (2006). *How picturebooks work*. New York: Routledge.
Oittinen, R., Ketola, A., & Garavini, M. (2018). *Translating picturebooks: Revoicing the verbal, the visual, and the aural for a child audience*. New York and London: Routledge.
Salisbury, M., & Styles, M. (2012). *Children's picturebooks: The art of visual storytelling*. London: Laurence King Publishing.
Thomas, P. (2000). *Trees: Their natural history*. Cambridge: Cambridge University Press.
Winters, S. F. (2018). Camping out on the quest: The landscape of boredom in *Harry Potter and the deathly hallows*. In A. Hudson (Ed.), *Children's literature and imaginative geography* (pp. 215–228). Waterloo, ON: Wilfrid Laurier University Press.

10 From Chamomiles to Oaks

Agency and Cultivation of Self-Awareness[1]

Andrea Casals Hill and Alida Mayne-Nicholls

Introduction

Gabriela Mistral's poem "La ronda de las manzanillas" (written c.1950, published in 1991) and the picturebook *El despertar del árbol* (2018) by Dídac P. Lagarriga, author, and Albert Asensio, illustrator, are texts meant for children featuring the agency of plants; chamomiles in Mistral's poem, and a young oak in the picturebook.[2] The poem and picturebook bring into play the creation of wholesome images and consistent worlds that affect the readers. Throughout her works, Mistral reveals a Franciscan consciousness, whereas Lagarriga and Asensio explicitly embrace posthumanist theory regarding the hidden life of trees (following Mancuso & Viola, 2016; Pigem, 2016; Wohlleben, 2016). Both texts feature plant and child agency and give a voice to the more-than-human world (Abram, 1996). This article seeks to analyse the characterization of plants in these two Iberoamerican texts building on childhood studies and plant studies; we are interested in the interspecies connection that both texts propose.

We focus on how they describe the natural and informal process of cultivating self-awareness, consciousness and the capacity of acting independently, diverting from traditional understandings of child development and linear evolution (James & James, 2012). Following Aloi (2018),

> The hope is that like human-animal studies, the field of plant studies will enrich our perspectives on plants, thus leading to different modalities in what right now constitutes a mostly unacknowledged critical node in the survival of life on the planet.
>
> (p. xxii)

That is, we hope that this essay contributes to the extension of Aloi's vision by discovering it in the work of creators who, in the past and in the present, have incorporated plants not as a mere decorative element, but as protagonists in their stories, as Haraway (2018) puts it, making "room for beings that do not fit into [conventional] cast[s] of characters" (p. 105).

DOI: 10.4324/9781032066356-14

Specifically, we demonstrate how both texts emphasize the relevance and the permanence of the communities of chamomiles and oaks. In the poem we find a community of chamomiles and a community of girls dancing the same *ronda*; in the picturebook we encounter a community of trees growing together and a community of children experiencing the forest. In both texts, knowledge and growth come from vivid experiences and interaction, and from "becoming-with each other" (Haraway 2018, p. 103).

Along with the analysis of the rhetorical and visual devices deployed by the authors and illustrator, we will ground our theoretical framework in Franciscan philosophy and childhood studies to expose the rich and complex images and constructions of Mistral's poem, and also in the fields of picturebook studies and posthumanism to unravel the subtleties in Lagarriga's text and Asensio's illustrations. First, we discuss the difficulties raised by the opposition between being and becoming, exposing the discrepancies between childhood studies and posthumanism, and propose a way to surpass this gap. Then we examine how oaks are constructed as active communicative subjects and chamomiles as lively collaborative dancers. We argue these texts posit plants as open to interspecies encounters, attuned with and aware of natural cycles, within a wholesome process that allows the reader to find a voice and an identity that is ontologically defined by experiencing nature (Gaard, 2008). Finally, we address how images conveyed by Mistral, Lagarriga and Asensio present cultivation of self-awareness as an experience intertwined with the process of being and becoming.

Being or Becoming

The expression "always becoming" is used to define basic notions in the posthumanist understanding of what it might mean to be human (Tarr & White, 2018). It opposes the Eurocentric standard of da Vinci's Vitruvian Man in as much as posthumanism understands that we are neither a fixed set of superlative characteristics, nor an exception among other beings (Schaeffer, 2009); nor are we separate or disconnected from our environment and the more-than-human world (Abram, 1996); nor are we evolving into a perfect human standard. Yet, in the intersection with childhood studies and children's cultural studies, these notions seem problematic as far as they bring back the idea of the child to the dubious figure of a human in process; that is, "always becoming" (Tarr & White, 2018, p. x); a vision of a future adult that cancels their present childhood. Childhood studies observe this as an issue of "futurity", in which children are "'becoming' rather than being" (James & James, 2012, p. 57). In a similar manner, we tend to value mature, full grown plants when they have *become* capable of producing fruits and timber or provide a sheltering shadow. Of course, girls and boys are developing and learning, but this adult-centric perspective deletes the notion of present and how children are being children, and rather establishes a notion of "arrival" (James & James, 2012, p. 58); a belief that at some point the becoming stages are completed and a full grown person appears. If we turn back to posthumanism, though, we acknowledge that adults are becoming as well; but childhood studies' view reminds us to consider children as they are in their *present being*, and restrain from thinking of them only in reference to their future adult forms.

The narrative proposed by Lagarriga's and Asensio's picturebook acknowledges the interconnectedness between the children and the trees, placing all of them in a vivid present, where the treeling is valued for its present being. Both, the narrative voice and the voice of the young tree are always enunciated in present tense. The large landscape format is environning (Casals, 2018), creating a reading ecosystem (Ramos & Ramos, 2011) that helps to draw the reader into that present time and place, and emphasizes the tree's perspective through the horizontal page spread in contrast to a human's perception of a tree, which would tend to be represented vertically because of the proportions of most trees. In the case of "manzanillas", Mistral (1991) builds an eternal present: the round is spinning now, a now projected from and into all time, "*desde todo tiempo*" (p. 138); that is, the round has been dancing forever in its *present being*. In this way, Mistral's poem captures the notion posited by childhood studies because it recognizes that girls are being girls in a lively present, and not becoming women.

The adult-centric perspective that childhood studies denounces challenges us, just as the anthropocentric perspective that posthumanism denounces does too. As we write these lines, we think of our own limited experience and our culturally biased reading. In *El despertar* we tend to concentrate on the depicted children's present and embodied experience amongst trees, rather than focusing on the young tree's embedded experience. Just as in "manzanillas" we tend to describe the dancing girls rather than the dancing chamomiles. This limitation springs from our own cultural background and "Westernised" anthropocentric thought, yet we recognize that the mediation fostered by the cover of the picturebook contributes to that anthropocentric reading (see Figure 10.1). Despite the book's title that could be translated as "The Tree's Awakening", which focuses on a tree, the girl with the red dress foregrounded sitting on a branch seems to demand the readers' attention (Kress & van Leeuwen, 2006). Following Aloi (2018), "Plant-blindness essentially is our cultural inability to conceive plants beyond the prefixed cultural schemata" (p. xx); thus, Asensio's unintentional error reduces the trees to a backdrop. From this visual paratext, readers anticipate a human protagonist that betrays the picturebook's intention to give voice to the trees and overcome the "post-anthropocentric view" that is at the heart of the posthumanist ideal (Herbrechter, 2013, p. 72).

As Lagarriga and Asensio explicitly state in the appendix of the book, they have been informed by Mancuso & Viola (2016), Pigem (2016), and Wohlleben (2016), and strongly recommend the reading of their texts which openly put forth posthumanist interrogations that question the notions that communication and intelligence are solely human, highlighting interconnectedness

Figure 10.1 Albert Asensio, illustrator. *El despertar del árbol.* Front cover. Rendered with permission.

and networks of collaborations amongst trees. As Tarr and White (2018) describe it: "Posthumanists deny the definition of humans as boundaries, exclusive, unique, exceptional, or naturally dominant. Instead, our intelligence, our bodies, our behaviour are interconnected with other species and the environment" (p. xi). Ahead of her time and any posthumanist figurations, Mistral already blurs boundaries and hierarchies between the human and the more-than-human world and puts forward non-hegemonic figures by giving voice to girls, single mothers, indigenous peoples, native fauna, and so on. Mistral initiates her "manzanillas" with the surprising voice "we", announcing an interconnectedness, which is underlined by the image of birds and other creatures dancing along with "us": "and they dance with us / birds and tiny beasts"[3](138).

Who is this "we" the poetic voice celebrates and constructs? Our cultural bias could lead us to think that the poetic voice is the collective voice of girls dancing a *ronda*, but as highlighted by the poem's title this round belongs to the chamomiles and not to the girls. As it happens in many of Mistral's rounds, the title introduces those who are dancing the *ronda* and summon others to join in. In this poem, chamomiles are the hosts calling the "*chiquillas*" (Mistral, 1991, p. 138) to spin with them (Mayne-Nicholls, 2018). These *chiquillas*, however, are not only human: birds and little beasts are included in the invitation. In this sense Mistral's "cast of characters" (Haraway, 2018, p. 105) are making kin with "other sorts of 'we', other sorts of 'selves' ... human and non human critters" (Haraway, 2018, p. 102). The vortex or climax of the Mistralian round is when dancers (chamomiles, girls, birds and so on) hold on to each other balancing centripetal and centrifugal forces, staying together as a "we" that spins as a whole. In Mistral's round we observe a clear isotopy regarding the feverish spinning of the round. In order to create *rondas* that are wild rather than restrained, Mistral includes words such as delirium, burning, fervour, crazy, and tremor (Mayne-Nicholls, 2018). In "manzanillas", all the *chiquillas* are immersed in this wild dance, in a movement that exudes individual and collective agency. There are two apparently contradictory situations going on simultaneously in the poem; though the reader can discern between chamomiles, girls and birds, the dancers are not just dancing for their own individual pleasure, but for the community, the sisterhood, where the "well-being" of one affects the other and impacts all (Braidotti following "Spinozist ethics", p. 340). Thus, there is interspecies connectedness and collaboration, and the collective affects the individuals just as each one's individuality is affected by being part of this collective ritual.

Although today we can discuss Mistral's poems from a posthumanist approach, hers is a Franciscan perspective. Mistral was a known follower of St Francis of Assisi, and it is possible to trace this in her writings.

> The key to an understanding of Francis is his belief in the virtue of humility—not merely for the individual but for a man as a species. Francis tried to depose man from his monarchy over creation and set up a democracy of all God's creatures.
>
> (White, 1996, p. 13)

We can notice this Franciscan humility and austerity even in Mistral herself, and how she chose to dress, from hair to toe;[4] but it is more eloquent in how she approaches life as a whole, in which humans are not the owners of the creation; an idea that is evident in her writing.

The relevance of the *ronda* is that no one is in a hierarchical position, just like there is no headboard in King Arthur's roundtable. Mistral goes beyond Arthur's arrangement: her *ronda* is not for male heroes, but for everyone, even the "damaged ones". In Mistral's round "Las que no danzan" ("The ones that don't dance"), the "damaged ones" are those who cannot dance: excluded people, animals, and flowers that seem ugly or useless; that is, in

Haraway's words, "the disregarded human persons and other displaced beings … that do not fit one's cast of characters" (2018, p. 105). At the same time, Mistral blurs the limits among species, and reunites them all in a Franciscan way:

> The Franciscan notion of nature collides with the idea of natural hierarchy between beings, establishing a more harmonious and solidary model amongst beings and all of nature. This model is sustained by all creatures' universal brotherhood; a notion that reveals Francis of Assisi's vision of nature as a whole, whose purpose is to reconcile and unify all of its beings. Thus, human beings are not above nature, rather we are with and within all nature.
>
> (Aguiló Bonet, 2009, p. 186)

The idea of being in and with all nature is featured by Mistral in the subjects she chose, and how they relate with each other, as we can see in "manzanillas". This *ronda* is summoned by the chamomiles that are reaching out for the girls and other creatures that the poetic voice reunites as a diverse group of peers: "with the *vicuña* and the alpaca, / with the *huemul* and the squirrel" (Mistral, 1991, p. 139). The poetic voice places together animals of different size and relevance from the Andean culture, disregarding their commercial value and stressing their merit as subjects in their own right. This is more evident in the tenth stanza, which is dedicated to the *chinchillas*. These animals were exploited for their fur around the 19th and 20th centuries, to the point of becoming an endangered species.[5] The poetic voice does not emphasize the beauty of the chinchillas's fur, but rather more subtle features: "tiny and quiet / and living slumberously" (Mistral, 1991, p. 139). In Mistral's imaginaries, all living creatures (human and the more-than-human) are equally important regardless of what humans can obtain from them.

Active Communicative Subjects

The connection and collaboration between species posited by posthumanist and Franciscan perspectives is evident in the rhetorical and visual devices deployed by the authors and illustrator to represent both chamomiles and oaks as active communicative subjects (Kokkola, 2017). That is the case of Lagarriga's and Asensio's picturebook. Though Barbara Bader's definition of picturebooks circulated over 40 years ago, her broad understanding is timely and can guide our appreciation of how the creators of the picturebook contribute to the construction of the treeling as an active communicative subject.

> A picturebook is text, illustrations, total design; an item of manufacture and a commercial product; a social, cultural, historical document; and

> foremost an experience for a child. As an art form it hinges on the interdependence of pictures and words, on the simultaneous display of two facing pages, and on the drama of turning pages.
>
> (Bader, 1976, p. 1)

Since her description, authors have tended to be more experimental and academics to regard picturebooks as postmodern cultural productions, particularly because of the polyphony of voices, or at least, the duality of voices offered by the picture/word interplay. Like poetry, picturebooks offer ambiguous meanings since they are open to interpretation, challenging the hyper-valued linear, objective and rational thought that privileges apparently unequivocal truths. Both ideas, the interdependence of pictures and words and their openness to multiple meanings make picturebooks a posthuman document themselves. A picturebook allows the reader to be immersed in the reading ecosystem it proposes (Ramos & Ramos, 2011), blurring the boundaries of the book (object) and the reader in the unique experience(s) it offers. From a neo-materialist point of view, the picturebook itself has agency as it transforms those who establish an intimate relation with it. Following Coole and Frost, Palacio (2018) asserts that acknowledging living matter's agency dismantles the myth that only humans have agency, and therefore deserve a privileged position to dominate and control nature (p. 11).

If we assume Bader's definition as the standard to value a picturebook, *El despertar* fully complies with these criteria. Most obviously, it tells a story by means of the interdependence between words and illustrations. The page spreads, the landscape design and the mesmerizing green create a reading ecosystem that environs the reader (Casals, 2018). Considering the picturebook as an entire design, even the endpapers in *El despertar* foster an immersive journey into the reading experience and into the depicted forest. The endpapers frame this experience inviting readers to closely observe "the hidden life of trees" (Wohlleben, 2016), to go beyond what meets the eye, literally looking under the surface, where much of the trees' communication and support takes place. The blurry brown watercolour from side to side on the lengthy horizontal spread conveys an immersive underground experience. The vision of the trees' roots reminds us of another poem by Mistral, "Raíces", where the poet embodies a transformative experience as she eagerly tries to learn what the roots hear; the poetic voice is underground, cuddled by the roots, her cheeks being caressed by them, as she says: "I yearn to know what they hear / to be as cuddled / I pass between them and my cheeks / are smudged with humid earth" (Mistral, 1967, p. 134).

Likewise, the endpapers in *El despertar* invite the young readers to look closely at all the details and learn to hear and see what the trees and their underground habitat communicate. Both endpapers depict the roots, but they are not the same: a mole and an ant are depicted on the front one; an ants' nest and laborious ants through their tunnels appear on the actual end page. The dissimilitude amongst the endpapers further invites the curious eye to observe nature closely and wonder. Framing the narrative with these paratexts helps contain the reader's experience within the narrative, welcoming the reader to experience the woods that are depicted, but also inviting the reader to go beyond the story, out there and observe.

Most outstandingly, *El despertar* functions as testimony of the times we are facing when the environmental crisis is a major concern, the "critical node in the survival of life on the planet" (Aloi, 2018, p. xxii); thus, the book becomes a tool for environmental education. The book teaches the reader to look closely into nature, to feel it, to hug trees, to notice life cycles, constructing a reader with a posthumanist thought, allowing the reader to find a voice and identity ontologically defined by experiencing nature (Gaard, 2008). It also gives testimony of the transition we are immersed in; the tension between the need and yearning to focus on, notice and care for the environment, and the limitations we have to actually free ourselves from our anthropocentric understanding of the world, which is made very obvious on the front cover discussed earlier. We cannot ignore that the book is an artefact meant for human children; "foremost an experience for a child" (Bader, 1976, p. 1), though, at the same time, it is meant to offer these young humans an immersive visual narrative from the perspective of a treeling. In that sense, Mistral's poem can also be considered a text to learn how to create community not just with those who are like us, but to overcome that anthropocentric approach and embrace a Franciscan point of view of an interspecies community. For example, the first two verses of Mistral's poem highlights that "the *ronda* that we like the most is that of chamomiles" (Mistral, 1991, p. 138) precisely because all species are invited to dance in this fraternal twist.

El despertar can be regarded as a posthumanist work because of the creators' explicit intention to direct the readers' attention to the imagined experience of the treeling, and because it is actually conveying this message in the form of a picturebook and the plurality of meaning this storytelling mode fosters (Barthes, 1982). That is, the multiplicity of meanings and interpretation, the subjectivity and ambiguity of the art form challenges the rational objective understanding of the world that sets humans as detached observers, inviting other ways of knowing. It is posthumanist too because the young tree's way of knowing the world around is through the senses: the young tree's experience is an embodied present, where he can feel and see the light, hear and feel the child's heartbeat as the children hug him.[6] On the first page spread,[7] the treeling sings: "I've only just awoken but I cannot see anything"; and goes on through the third spread: "Mama, can you hear me?". The bigger trees also perceive and communicate through their senses: "a big tree perceives the scent emitted by the treeling. / They chat through the flowers that are just sprouting." On the fourth spread the visual focus is on the branches of the treeling that children are climbing. Text and image are consistent in drawing the readers' attention to the tree's experience: "The treeling also notices the wind that caresses him / and the weight of the children who, climbing on his branches, look closely at the lines of ants".

As a whole, the visual narrative presents a cyclical and circular experience. The picturebook opens and ends in the early spring, passing through the four seasons, adding another posthumanist perspective since it is not a linear story with a clear goal; not a traditional narrative with rising action, climax and denouement; thus, shifting the anthropocentric notion that literary works are all about "human concern" (Gibson, 2009, p. 483), positioning the reader in the treeling's perspective as an actant rather than a passive character, as Aloi (2018) suggests:

> At stake is the opportunity to understand plants as integral, coexisting *actants* that play defining roles in the functioning of ecosystems on this planet. What we look at, and how we look, constitute essential parameters in the recuperation of "alternative gazes" and the crafting of new ones—modalities of engagement that entail more than the ocular—modalities that can lead to a reontologization of the living.
>
> (p. xx)

It seems evident that in their appreciation of the pedagogical function of children's literature Lagarriga and Asensio see the defining role of the treeling's story told from his perspective as an opportunity to understand plants as integral coexisting actants. In their picturebook, they focus on the treeling's alternative gaze that is not only ocular; the treeling enjoys the children's embrace and beating of their heart:

> Mama, what's this. I like it—says the treeling ... —And those soft beats on my bark, what are they? ... [the children] are so close to the trees that they notice how their tiny hearts beat happily and, at the same time, the children receive from the trees all the strength, energy and serenity that emanates from them.
>
> (8th, 9th, 10th page spreads)

Thus, *El despertar* allows its young readers to engage with the ecosystem using all of their senses as well as considering an alternative gaze, and even being perceived by other critters.

Lively Collaborative Dancers

Mistral wrote children's poetry all along her professional career. She was certain that children needed to read poetry specially dedicated to them, but not poems conceived in didactic terms: "I have wanted to write a new school poetry, because the one in vogue does not satisfy me; a school poetry that does not stop being poetry because it is school" (Mistral, 2013, p. 49). When Mistral started writing her verses on childhood, around the 1910s and 1920s, children's poetry in Spain and Latin America had an instructive and moralizing character (García Padrino, 2010; Rodríguez, 2010). For Mistral

(1935), that kind of poetry was "babbles of teachers" (p. 145), whereas she was concerned with stressing the aesthetic experience of reading poetry. The result is a poetry that *appears* to be simple on a heuristic level. We can observe this in some of the rhetorical devices deployed by Mistral. For example, "manzanillas" is formed by 13 stanzas, most of them composed of four verses (the second and seventh stanzas have six verses). The verses are octosyllabic, except the last one that has 11 syllables. This structure seems to adjust to the *rondeau* and the *rondel* developed by French poets in the Middle Ages (Baldick, 2008). These poems had a strict format regarding metrics, number of verses and stanzas, and type of rhyme. In "manzanillas" Mistral uses the same number of stanzas as *rondeaus*, but she grouped them differently. Regarding the rhyme, she uses both consonant and assonant rhymes.

Mistral wrote in a period in which poets were exploring and advancing Modernism (the older poets) and Avant-Gardism (the younger poets). Mistral was in an interstitium that allowed her to develop a poetry of her own. In that sense, Mistral used these very structured formats, but the content, the words chosen, offer a counterpoint as Mistral posited new images and ways of approaching life. In "manzanillas" Mistral elaborated on this structure with sonorous vernacular words, everyday language, and colloquial idiomatic expressions that place the readers in Montegrande, her hometown. This is especially notorious in her use of consonant rhyme in the first three stanzas: *manzanillas, villas, chiquillas, ambiciosilla, bestecillas, maravilla* [8] (Mistral, 1991, p. 138).[9] The suffix "*-illa*" (/ˈija/) can indicate a diminutive or an expression of affection. That is the case of *chiquillas, ambiciosilla* and *bestecillas*, that introduce the reader in a world of small creatures such as chamomiles. The poetic voice is presenting the creatures native to Montegrande which the poet knows well because they inhabit and share the same place where she grew up. If the point of view in the poem is that of the camomile, then the tiny wildflower is establishing the dimensions: little ones are peers, presenting their own world.

Mistral also proposed that children's poetry should consider the local folklore, that is, the use of familiar images and familiar words; and the construction of verses that seem spontaneous, that enjoy musicality and rhythm (Mayne-Nicholls, 2018). Accordingly, Mistral incorporates this vernacular insistence to "manzanillas" in the use of diminutives to exude a mood that is fresh and vibrant; she could have chosen other forms of diminutives (like *chiquitas*), but the ending *-illa* provides musicality to the verse and a tenderness that the *-ita* ending does not evoke because of the strong plosive. That lively mood is consistent with the feverish energy of the round.

Dancing along a *ronda* implies movement in a circle in motion. In "manzanillas" Mistral emphasizes this condition; in fact, her *ronda* is a continuous and elated motion. When theorizing about rounds, Mistral presents an analogy between children and witches. She describes children like witches are often imagined in popular culture; that is, dancing around the fireplace in a coven where they become one. The bond[10] between the dancers and the colour of their faces are fiery, both among the children and among the witches (Mistral, n. d.). The dance she figures in "manzanillas" is fast and energetic; Mistral transmits this by using action verbs—*spin, dance, rise, return, enlarge*—throughout the poem, so the reader has the impression that the round never stops. This idea is stressed by confronting contrary states: at one point the round begins as a small one, and then enlarges; or it was barely heard, and then it sounds like a loud living thing; or some of the dancers were napping and then they were awoken by the surprising collective movement. The intensity increases towards the seventh stanza, when a poetic subject—Mistral herself—is presented: "dancing in my Montegrande / as a lost madwoman" (Mistral, 1991, p. 139). The presence of the "lost madwoman"[11] represents a significant moment of the dance, a point of no return in which the braided dancers are so immersed in the *ronda*'s feverish movement, that it cannot be stopped, so the only next possible step is to ascend altogether. Ascension is a recurrent image in Mistral's rounds in which the centrifugal force generated by the rapturous turn of the dancers causes the round to rise. In "manzanillas", this moment coincides with the presence of Mother Mary who dances in heaven, as we can see in the tenth stanza: "and it happens that any day to heaven they ascend to give round to Virgin Mary" (Mistral, 1991, p. 139). Hierarchies have been subverted and therefore these tiny beings can even ascend to dance in heaven.

In a performative and transgressive turn, the poem erases boundaries: like the poet joining the transformative spin, like the creatures awoken and empowered in the ascending communal whirl, the readers are transformed as they read the expanding and soaring verses. We readers, chamomiles, tiny beasts, girls and poet, all become one in the performative act of reading and dancing along. In an avant-garde composition, Mistral has erased hierarchies and boundaries between species and also between fiction and reality.

We propose that the persona in this poem is the chamomile, because she summons all the dancers onto the round and because of its form. That is, Mistral choses chamomiles for their wilderness, their size and form. Chamomile flowers are circular and contain in themselves a round composed of petals tightly arranged together like the girls holding their hands in the *ronda*. Mistral (1945) builds the same image in other poems around the figure of the sunflower ("La pajita"), which projects like a gigantic chamomile. Aloi (2018) argues that "Plants' fixity, perceived passivity, and resilient silent presence have, for over two thousand years relegated plants to cultural backgrounds" (p. xxiii). We do not see that in Mistral's poem; chamomiles are neither fixed nor passive, and they are certainly not a backdrop for dancing girls. On the contrary, they emerge as the measure of things defining the composition of the poem.

The rounds proposed by Lagarriga and Asensio have a different rhythm from those offered by Mistral. They are gentle and contemplative, and together with the autumn setting (see Figures 10.1–3), they invite readers to an inward gesture, rather than Mistral's ascending frenzy. As the children hug the trees, the treeling perceives the children's hearts beating on his bark.

Figure 10.2 Albert Asensio, illustrator. *El despertar del árbol*. Page spread 10. Rendered with permission.

Figure 10.3 Albert Asensio, illustrator. *El despertar del árbol*. Page spread 13. Rendered with permission.

The illustration of the children laying in a circle on the ground that is covered with leaves is the only detailed image of their faces, yet the perspective is that of the tree looking at them from above. This image is complemented by the narrative voice explaining that the children collect dry leaves and listen to them as if their sounds were the trees telling what they have seen. Though the illustration seems to foreground the humans, the viewer here is the tree and it is his *gaze* as an *actant* that gives meaning to the spread. The large dimensions of the book, the landscape layout of the spreads, and the lack of apparent conflict and climax, invite a slow reading that echoes the slow growth of the trees within a year's cycle that the book reproduces by ending almost with the same scene it began with: the treeling asking "Mama, do you hear me?"; yet, an attentive reader will notice that the tree is thicker and taller.

Conclusions: To the Cultivation of Self-Awareness

In the poem and the picturebook, we find no abuse of nature. The local forest in the picturebook and endemic beings in the poem are at the centre, challenging anthropocentric attitudes and notions of productivity: children experience the forest in a leisurely manner, just as chamomiles, girls and other creatures spin round merrily, all embodied and present. In both texts there is an active community and an interspecies exchange and collaboration, creating "assemblages of human and non-human actors" (Braidotti, 2018, p. 339). In a Franciscan turn, in this assemblage there are no hierarchies, and therefore no one is excluded.

Aloi called our inability to look at plants as vibrant beings "plant-blindness". This cultural prejudice "reduces [plants] to resources or aesthetic objects" (Aloi, 2018, p. xx). However, neither the chamomiles nor the oaks in these texts are ornaments or resources. The contrast between them provides a perspective that reveals that within their own imagined universe, both the chamomile and the oak are the measure of things. The lack of fuss, the steady slow rhythm in *El despertar* is a consistent structure to portray the growth of a young oak, subtly mirroring a distinctive feature and time scale of large trees. In an apparent contradiction, Mistral's feverish *ronda* seems attuned with the fast growth of a tiny wild flower like a chamomile; thus, the ronda spins in excitement for the chamomile's burst of life.

The environing features of the picturebook and performative quality of the poem allow the readers to experience the slow rhythm of the oak as well as the fiery rhythm of the chamomiles, which awakens the readers, as the treeling becomes aware of himself as an embedded member of his community, and as those slumbering in Mistral's round awaken to the dance. The treeling's sensorial awareness and the chamomiles' performative dance of becoming one with all give rise to the readers' own interconnected awareness; when in *El despertar* the children touch each other and the oak while making a round as they hug the tree, or in the poem when all creatures become one in the fiery dance. Thus, the children reading, the children depicted in the forest, and the human dancers in the round become active participants with and within the vivid and sentient nature represented in these texts. Considering *El despertar*, it seems almost redundant to notice that this story revolves around the idea of the treeling's *awakening* to a new cycle of becomings amongst his community, which, at the same time, mediates the readers' awareness both of the treeling's experience, as well as their possibility of becoming entangled in making kin. In Mistral's *ronda*, by waking up to join the round and getting carried away by the dance, girls become girls, understanding who they are and what they can achieve as participants in this interspecies dance. This is explicitly exposed by Mistral (1991): "And those who napped / wake up surprised" (p. 138). The amazement of feeling awake and of being part of something bigger than oneself is a constant theme in Mistral's rounds, in which the never-ending movement prompts awareness of life as a continuous motion and being part of that motion implies a connection between every single being.

We are conscious that the poem and the picturebook analysed are not an obvious corpus to work with but taking on the challenge to explore through this unusual selection has allowed us to broaden our own gaze and "enrich our perspectives on plants" (Aloi, 2018, p. xxii), and discover this viewpoint not only in recent texts, but also in much earlier works. By bringing together these texts, we presented two different stages or moments. In *El despertar* we see those first crucial, direct childhood reciprocal encounters: the treeling and the children are awakening together; in "manzanillas", chamomiles are so aware of this awakening, that they summon others to join the vibrant dance of life.

Notes

1 This article is part of Mayne-Nicholls' Doctoral research "Espacios tomados: Representación de las niñas en Gabriela Mistral y María Flora Yáñez" (*Beca Doctorado Nacional Conicyt* #2113063); and Casals' postdoctoral research "Conciencia ambiental en la literatura y escritura infantil a juvenil ilustrada chilena del nuevo milenio" (Fondecyt-Conicyt #3170134).

2 Our somewhat dissimilar literary corpus in terms of genre and publication dates allows an intellectual challenge that represents a multimodal conversation in the spirit of "thinking-with" (Haraway, 2018, p. 103) and "creating connections where things were previously disconnected or seemed unrelated" (Lykke, 2018, p. 333).

3 All translations from Spanish to English are ours.

4 "[Mistral] tried to disarm the hegemonic constructions on women, again through her work and bringing this to her own body … in the way she dressed or combed her hair … Mistral does not dress to be a lady of society, according to the customs of the time, but to perform as a woman writer" (Mayne-Nicholls, 2018, p. 37).

5 In her poem "La chinchilla", Mistral denounces that chinchillas are desired and harassed by huntsmen.

6 We respect the masculine form expressed in the original.

7 Given that *El despertar* lacks page numbers, we reference the quotes in the order of the page spreads.

8 Here, the word *maravilla* refers to wonder, yet, in Chilean Spanish, it also means sunflower.

9 In the rest of the poem there are other words ending in *-illa*, such as *trilla, ardilla* and *chinchilla.*
10 An isotopy present in many of Mistral's *rondas* is the braid. The hands of the girls are entwined; that is, their fingers are interspersed as threads of a braid. This condition makes the round strong even with the rapturous spinning (Mayne-Nicholls, 2018).
11 The idea of the "loca perdida" is recurrent in Mistral's poetry.

References

Abram, D. (1996). *Spell of the sensuous.* New York: Vintage Books.

Aguiló Bonet, A. J. (2009). Actitudes antropocéntricas y biocéntricas en las teologías judeocristianas de la naturaleza: una aproximación crítica. *Ludus Vitalis*, XVII (31), 171–198.

Aloi, G. (2018). *Why look at plants? The botanical emergence in contemporary art.* Boston, MA: Brill.

Bader, B. (1976). *American picturebooks from Noah's ark to the beast within.* New York: Macmillan.

Baldick, C. (2008). *Oxford dictionary of literary terms* (3rd ed.). Oxford: Oxford University Press.

Barthes, R. (1982). *El placer del texto & Lección inaugural* (N. Rosa & O. Terán, Trans.). Ciudad de México, México: Siglo XXI.

Braidotti, R. (2018). Posthuman critical theory. In R. Braidotti & M. Hlavajova (Eds.), *Posthuman glossary* (pp. 339–342). London: Bloomsbury.

Casals, A. (2018). Hacia una ética ecológica en la literatura infantil a juvenil chilena del nuevo milenio. *Anales de Literatura Chilena*, 30, 257–271.

Gaard, G. (2008). Toward an ecopedagogy of children's environmental literature. *Green Theory & Praxis: The Journal of Ecopedagogy*, 4(2). https://doi.org/10.3903/gtp.2008.2.3

García Padrino, J. (2010). La literatura infantil y juvenil en España. Siglos XIX y XX, hasta 1980. In *Actas y memorias del Congreso Iberoamericano de Lengua y Literatura Infantil y Juvenil* (pp. 43–65). Santiago, Chile: Fundación SM.

Gibson, J. (2009). Literature and knowledge. In R. Eldridge (Ed.), *The Oxford handbook of philosophy and literature* (pp. 467–468). Oxford: Oxford University Press.

Haraway, D. (2018). Staying with trouble for multispecies environmental justice. *Dialogues in Human Geography*, 8(1), 102–105.

Herbrechter, S. (2013). *Posthumanism: A critical analysis* [Kindle Voyage]. London: Bloomsbury.

James, A., & James, A. (2012). *Key concepts in childhood studies.* London: Sage.

Kokkola, L. (2017). Critical plant studies and children's literature. In C. Beauvais & M. Nikolajeva (Eds.), *The Edinburgh companion to children's literature* (pp. 274–280). Edinburgh: Edinburgh University Press.

Kress, G., & van Leeuwen, T. (2006). *Reading images: A grammar of visual design.* London: Routledge.

Lagarriga, D., & Asensio, A. (2018). *El despertar del árbol.* Barcelona: Akiara Books.

Lykke, N. (2018). Postdisciplinarity. In R. Braidotti & M. Hlavajova (Eds.), *Posthuman glossary* (pp. 332–335). London: Bloomsbury.

Mancuso, S., & Viola, A. (2016). *Sensibilidad e inteligencia en el mundo vegetal* (D. Paradela López, Trans.). Barcelona: Galaxia Gutenberg.

Mayne-Nicholls, A. (2018). *Espacios tomados: Representación de las niñas en Gabriela Mistral y María Flora Yáñez*. (Doctoral diss.).

Mistral, G. (1935). El folklore para los niños. *Revista de Pedagogía*, 160, 145–152.

Mistral, G. (1945). La pajita; La ronda del arco-iris; Ronda de los metales. *Ternura*. Buenos Aires: Espasa Calpe.

Mistral, G. (1967). La chinchilla; Raíces. *Poema de Chile*. Santiago, Chile: Pomaire.

Mistral, G. (1991). Ronda de las manzanillas. *Lagar* II. Santiago, Chile: Ediciones de la Dirección de Bibliotecas, Archivos y Museos.

Mistral, G. (2013). *Vivir y escribir* (P. P. Zegers, Ed.). Santiago, Chile: Ediciones Universidad Diego Portales.

Mistral, G. (n. d.). Rondas [manuscript]. Archivo del Escritor.

Palacio, M. (2018). Neo-materialismo: ¿un retorno de la metafísica en la nueva filosofía de la naturaleza? In M. Palacio (Ed.), *Neo-materialismo. La vida humana, la materia viviente y el cosmos* (pp. 9–25). Buenos Aires: Prometeo.

Pigem, J. (2016). *Inteligencia vital*. Barcelona: Kairós.

Ramos, A. M., & Ramos, R. (2011). Ecoliteracy through imagery: A close reading of two wordless picture books. *Children's Literature in Education*, 42, 325–339.

Rodríguez, A. O. (2010). Literatura infantil y juvenil latinoamericana. De los orígenes a los años 70: algunos hitos y tendencias. In *Actas y memorias del Congreso Iberoamericano de Lengua y Literatura Infantil y Juvenil* (pp. 66–79). Santiago, Chile: Fundación SM.

Schaeffer, J. M. (2009). *El fin de la excepción humana*. Buenos Aires: Fondo de Cultura Económica.

Tarr, A., & White, D. R. (2018). *Posthumanism in young adult fiction: Finding humanity in a posthuman world*. Jackson: University of Mississippi Press.

White, Jr., L. (1996). The historical roots of our ecologic crisis. In C. Glotfelty & H. Fromm (Eds.), *The ecocriticism reader* (pp. 3–14). Athens, GA: The University of Georgia Press.

Wohlleben, P. (2016). *The hidden life of trees: What they feel, how they communicate. Discoveries from a secret world* (J. Billinghurst, Trans.). Vancouver: Greystone Books.

Part IV

Plant Agency and Activism

11 Vegetal Individuals and Plant Agency in Twenty-First Century Children's Literature

Anja Höing

Having long been standard features in children's literature, representations of plants have recently gained new prominence due to a significant shift in their narrative function: rather than structuring the setting or operating as symbols, highly individualized and often agentic plants now people the character cast of children's literature. Yet, vegetal agency has as yet hardly, if ever, been at the centre of attention in approaches to agency in twenty-first century children's literature, nor has children's literature been a central focal point in approaches to representations of plant agency. Research on agency in children's literature focuses on the agentic capacities of the child itself, often with regard to gender and diversity, and most centrally in juxtaposition or conflict with adult agency and society structures and institutions (Clark & Castro, 2019). Plant agency, in contrast, is most frequently analysed in literary or cinematic environments in which agentic plants are antagonistic or even monstrous, i.e. in the Gothic tradition and contemporary horror (e.g. Keetley, 2016). Bringing together perspectives from both these fields of research, this chapter will explore the representation of vegetal protagonists in three recently published English-language children's books: Kathleen Van Cleve's *Drizzle* (2010), Frances Hardinge's *The Lie Tree* (2015), and Sophie Anderson's *The Girl Who Speaks Bear* (2019). Hardinge and Anderson are two of contemporary Britain's most productive bestselling authors whose works draw huge readerships. Yet they focus on different target groups: while Hardinge primarily writes for young adults, or children nearing adolescence, Anderson's stories address a younger readership. Van Cleve is not as well-known as these two British authors, but *Drizzle*, her first foray into children's literature, stands out as one of very few US-American texts to participate in what to date appears to be primarily a British or European writing tradition of interrogating child–plant relationships. In each of the three narratives, a child protagonist interacts with an individual agentic plant in a semi-fantastic setting, yet, the three narratives sketch vastly different visions of child–vegetal interaction. While young Polly in Van Cleve's *Drizzle* experiences a deep friendship with the intelligent rhubarb plant Harry, Hardinge's protagonist Faith has to disentangle herself from the clutches of a vegetal false friend, the both disturbing

DOI: 10.4324/9781032066356-16

and profoundly dangerous Lie Tree. An intelligent tree is also at the heart of *The Girl Who Speaks Bear*, as bear-human hybrid Yanka embarks on a quest to save the magical lime tree that governs natural order itself. Taking together the vastly different plant–child interactions set out in these children's books, this chapter will investigate how such fictional representations of plant–child interaction do on the one hand challenge traditional readings of plants in literature and suggest new ways of making sense of human–plant ties, but, contradictorily, simultaneously reinforce exactly those traditional readings that block any potential of understanding a vegetal individual as something other than a metaphorical extension of, or instrument in the realization of the human Self.

In order to outline the parameters governing the distinction this chapter draws between traditional and new ways of representing literary plants, a short excursion into the ecocritical framework that informs this analysis is necessary. Two central and interconnected ecocritical issues are of particular importance to this chapter: the dualism of natural vs. unnatural and the myth of a static nature. As for example William Cronon and Kate Soper have pointed out, in British or Anglo-American cultural frameworks humans—and hence the cultural representations they produce—tend to construct dichotomical either/or frameworks to differentiate what is "natural" from what is not (Cronon, 1996, p. 36; Soper, 2015, p. 275). As Ecofeminist scholar Val Plumwood has famously argued, such dualisms are not only problematic in that they construct stark borders where there are none; they also impose implicit hierarchies and ideological framings on the concepts thus set in opposition to each other (Plumwood, 1993, p. 41). With regard to the natural/unnatural dualism, anything branded as "unnatural" is simultaneously constructed as morally suspect. As Cronon phrases the issue, "the natural" has the "capacity to take disputed values and make them seem innate, essential, eternal, nonnegotiable", while simultaneously ostracizing any alternative position as "unnatural" (Cronon, 1996, p. 36). What is more, "the natural" is often associated with stasis—ironically so, given that scientifically the concept of a static nature has been discarded for centuries and contemporary ecology perceives of natural systems as adaptive and changeable (Höing, 2017, pp. 4–5). Nonetheless, the conception of a static nature continues to dominate the environmental imagination of many people until today (Garrard, 2012, p. 63; Cronon 1996, p. 24), and the concurrent misunderstandings of basic parameters of natural systems are frequently at the root of problems environmental activism finds itself pitted against. Plants, and especially trees, might even buttress illusions of nature as static. As Ryan points out, we tend to "think botanical life is sessile (unmoving), silent (lacking address), passive (acted upon by mobile life forms), and … pleasing (agreeable to the senses)" (Ryan, 2018, p. 1). Tree movements in particular indeed often happen in time frames our eyes cannot perceive, and their life span is so long that humans tend to classify them as part of the setting rather than as living and breathing individuals. For this chapter, this theoretical framework provides two central axes which the three children's stories will be read against: Is vegetal agency and child–plant interaction constructed as natural or as unnatural, and is the framework on which vegetal agency is acted out constructed as a static or as a changeable, adaptive one? As this chapter will argue, all three texts—though in vastly different ways—graft their visions of vegetal agency on a stark natural vs. unnatural dualism as well as on conceptions of a static nature. Despite all instances of vegetal emancipation the three books experiment with, in such static, dualist networks the central position of the human Self in a thoroughly anthropocentric world remains ultimately unchallenged.

Van Cleve's *Drizzle:* Vegetal–Human Friendship and Self-Realization

In a hopeful, upbeat vision of human–vegetal interaction Kathleen Van Cleve's *Drizzle* (2010) represents plants and humans as virtually symbiotic partners in an animate and even magical ecological system—but ultimately the story still resolves around humans, and its conclusion reinforces the very anthropocentric structures the earlier chapters ostensibly break down. *Drizzle* is primarily concerned with two topics addressed in many works of children's literature: the value of friendship, and self-realization. In that, the story is not unusual, one could even argue that it follows a standard pattern. The heroine, Polly, feeling different from other children and hence struggling to socialize into her peer group, has to learn to trust into her true friends in order to gain the necessary self-confidence that will enable her to uncover her unique powers, which in turn will save both her family's magical rhubarb farm and her own status in her peer group.

Drizzle becomes unusual once one narrows the gaze to the nature of Polly's companions. Polly's best friend is vegetal. Polly, acting as homodiegetic narrator relating the story in diary-fashion, introduces the agentic plant at an early point in the story:

> I have to tell my best friend, Harry … Usually he has some good ideas. Mom would like him, if she ever met him.
>
> Actually, she has met him—she just doesn't know it. He's always on the farm, west field, row eighteen, column thirty. This is because Harry, my best friend, is a chocolate rhubarb plant.
>
> (2010, p. 8)

Polly introduces Harry as first, her best friend, a gendered individual with a conventional human name, second, as very intelligent and likeable, and only third as vegetal. This structure clearly aims to surprise the reader, and in doing so arguably assumes that the implied reader will not expect, or perhaps not even deem possible the implied human–vegetal friendship. This is reinforced a second time in the repetitive parenthesis "my best friend" in the final sentence of Polly's introduction to Harry. The issue that Harry is not only a named vegetal, but, being this, can also be a best friend, apparently needs repetition to be driven home.

Despite such clear markers that signpost the human–vegetal friendship between Harry and Polly as counter-intuitive, the narrative attaches high value to their connection. Even when Polly finds a human friend, the text does not present it as necessary that she sacrifices her vegetal friend in pursuit of a more "valuable" intra-species friendship. One can read this development in direct opposition to many late twentieth-century representations of inter-species friendships in children's literature, in which a disconnection from a non-human companion, or even the companion's death, is represented as a necessary step in a child's maturation process (Johnson 2000; Hogan 2009). Harry and Polly's interspecies friendship in *Drizzle*, in contrast, follows a pattern more common to children's books narrating intra-species friendships: the two quarrel severely, even fight, and finally make up. The story of their friendship ends as it began, with Polly again "go[ing] out there every morning, telling [Harry] everything" (2010, p. 356). Although Polly gains human friends too, she gains them in addition, not instead of her vegetal friendship, and a crucial test for her true human friend is his willingness to accept the rhubarb as an equal as regards Polly's affections (p. 102).

In many ways, *Drizzle* destabilizes traditional readings of vegetal agency. Human–plant communication—and the failure thereof—is a major topic in the book. When introducing Harry, Polly explains to the reader:

> Obviously Harry can't talk. He's a plant, not a person. Still, he communicates—all plants communicate. And not just on 'magic' farms. It happens *everywhere*. If you don't believe it … it's because you're not really paying attention. I'm not saying they're going to *talk* back. Even I know that would be crazy. But they'll answer you. It's a fact.
>
> (p. 8, emphasis in original)

The narrator's fervent plea towards acknowledging vegetal agency directly targets the implied reader, who, unlike Polly, does not live in a "magic" place, but should still pay attention to plant communication. Later in the plotline, the reader will also witness the fatal consequences following a breakdown of human–vegetal communication in the story world. When Polly repeatedly fails to understand an important message Harry tries to convey, she vents her frustration by severely wounding the rhubarb (p. 115). This has direct consequences for the individual—for the rest of the story a humbled, repentant Polly will nurse her best friend back to health. But more importantly, Polly's own failure to "[pay] attention" (p. 8) endangers the continued existence of the entire "ecosystem" the story is set in: her family's farm, whose desertification only she can counteract. Polly's development throughout the story clearly proposes a role model for the implied reader: in addition to conceding that she, not Harry is to blame for the breakdown of human–vegetal communication, she needs to acknowledge that the entire farm is a complex network of animate nature that spans not only the crop of rhubarbs but also the dragonflies over the lake, a "truly weeping cherry blossom tree" (p. 2), and the slugs she abhors. Only when she has learned all these lessons, she can become an ecological agent and successfully battle the local climate change that endangers the farm.

Yet, in other respects, the representation of human–plant relationships in *Drizzle* is deeply, even disturbingly traditional. Ultimately, the plants are enablers only. They act as merely a part of a system whose central axis is an anthropocentric one. At the conclusion of the story the reader learns that the entire, ultimately static ecosystem of the rhubarb farm only persists because magical human females such as Polly can call rain. Without these humans in the centre, the system would collapse. The ecosystem of the farm is not, to make use of a deeply ideologically loaded term, "natural". In consequence, the climate change Polly combats in the story is also of an altogether different breed than the one in the real world: Polly's discovery of her agency serves to keep stable a static "magical" system—the desertification that threatens the place would merely be the farm reverting to its "natural" state. In the end, the entire system Polly lives in is exposed as an artificial one. The farm is a virtually hydroponic environment solely dependent on, and ultimately created for, human beings who quite tellingly also exploit its extraordinariness as a theme park—which the narrative does neither construct as problematic, nor as in any way paradoxical. The artificiality of the system that embowers Polly's environmental agency renders it fairy-tale-like, and implicitly counteracts the narrator's explicit plea for acknowledging vegetal agency even outside its boundaries. Such wonderful agentic plants humans can not only meaningfully communicate, but truly bond with, the reader might conclude, are as "natural" as the ecological system of the farm. Van Cleve's supernatural settings render their individual environment a magic and happier place, but ultimately an impossible one.

Hence, one might argue that *Drizzle* in its representation of the agentic rhubarb Harry and his environment ultimately constructs a paradox: while it indeed deliberately subverts some myths that render traditional plant representation problematic, it covertly reinforces others. Ultimately the book falls back on retelling one of the oldest narratives of children's literature—the issue of self-realization (reinforced through repeated intertextual reference to Emerson's *Self Reliance*) in a slightly green-washed fashion. In the end, all the plants' agency condenses into a single focal point: Polly has to explore her own identity, and to discover her Self—through agentic plants that act as vessel.

Frances Hardinge's *The Lie Tree*: The Agentic Plant as Gothic Other

In many ways, the vision of vegetal agency in Frances Hardinge's *The Lie Tree* (2015) is diametrically opposed to the one in *Drizzle*, yet, as I will argue, there are crucial similarities in that in Hardinge's story, too, vegetal agency ultimately serves to empower a human agent, and in itself is coded as "unnatural". The relationship between this human agent and the vegetal one is drastically different from the one in *Drizzle*. Hardinge's agentic plant can firmly be situated in the literary realm that to this day has given birth to most agentic plants in the English literary tradition: the Gothic, and its contemporary successor, the horror story. Vegetal agency bears a lingering fascination especially for the (eco-)Gothic. As Keetley argues in an essay proposing six theses on plant horror (2016), one reason for this might be that an entity that is simultaneously "[rooted] in place" and capable of significant movement (p. 13) is frightening because humans tend to perceive vegetal growth, which is so different from animal movement, as necessarily transgressive, even as "monstrous" (p. 13). Keetley traces a tradition of representing horrifying plants as a "multiplying mass" (p. 14), which finds a further example in Hardinge's narrative. Hardinge's lie tree grows throughout the story, from a "small tangle of blackened, dead-looking fines, stripped of foliage" (2015, p. 168) to "a mass of writhing black creeper, so dense and dark that it looked like a portal into an abyss" (p. 391), and that on the cover page threateningly bursts the confinements of its pot. Additionally, the plant has vampiric features. Dracula-like the tree renders Hardinge's protagonist Faith powerless to resist its influence, feeds on her lies, and draws her life power; it even has her suck its own blood in form of its fruit. Quite tellingly, the plant has to be burned in order for Faith to fully escape its devastating influence. This plant's agency is primarily unsettling, and adds profoundly to the dark and disturbing atmosphere of the book. The tree is a menace that would have better been left alone. Faith proves herself to be the only one who can disentangle herself from its influence, albeit stripped of even the last remnants of a childhood innocence that was precarious from the beginning.

Even more strongly than in *Drizzle*, where the plants also contribute to the perpetuation of an artificial system, the Lie Tree is associated with the unnatural. Not only does the tree thrive in a place no real plant would survive in—in total darkness—its main function proves to be to hinder human realization of how natural systems actually function. By supporting Faith's father in his self-imposed mission of fabricating fake fossils to keep Darwin's evolutionary theory from catching on, the Lie Tree is complicit in keeping up an illusion of stasis in face of a changeable and continually evolving reality. There is a strong network of dark irony at work in *The Lie Tree*—not only in the name of Faith, who after exposing her father's deception, will head off into the world as a figurehead of the evolutionary theory (Hardinge, 2015, p. 410), but also in that the plant itself, unnatural as it may be, in a way only strives for its Darwinian survival—and quite tellingly finds its end in a chapter entitled "Survival and the Fittest". In a world *really* governed by the system Darwin discovered, this plant would not be able to exist. Its realm is the wondrous one of the static intelligent design Faith's father believes in, and, much like the fairy Tinkerbell in Peter Pan that will perish once children stop believing in her, the tree dies with Faith's belief in the system it is rooted in. The first step towards its inevitable end is already taken in its "Christening": referring to a plant that several of the antagonists believe to be the biblical "Tree of

Knowledge" (pp. 389–390) as the "Lie Tree" destabilizes the entire "regime of truth", to borrow from Foucault (2002, p. 131), this alleged "knowledge" is founded on.

The Lie Tree hence transmits a carefully designed ideological message: advocating science over religion, progress over stasis,[1] and agentic self-determination over submission to structure. Yet, with regard to this ideological background, vegetal agency finds itself somewhat curiously placed on what necessarily has to be the losing side. The story thrives on the dualist oppositions set out above, and the Lie Tree is associated with the villains and the dangerously outdated world view they seek to preserve at all costs. Vegetal agency in consequence is coded as unnatural as the belief system it symbioses with. There is no place for it in Faith's newly discovered world of rational science. As are Van Cleve's magic rhubarbs, the Lie Tree is cast into the world of fairy tales—yet forced into the role of the antagonist. One could even finish a reading of *The Lie Tree* convinced that lies and agentic plants are one and the same thing: evil unnatural entities that have to be dragged into the light and destroyed for the greater good, namely the advent of science. The role of vegetal agency in this vision is rather dispiriting. In a comprehensive triumph of human exceptionalism over vegetal agency, one of Faith's final outbursts when facing the Lie Tree is: "It is only a *plant*" (Hardinge, 2015, p. 394, original emphasis). The insubordinate vegetal is put in its place in a system in which inferiority is obviously considered an essential quality of being vegetal. All this "mere" vegetal's mysterious agentic capacities are soon drowned out by the absolute victory of a scientific discourse that has its own ways of evading uncomfortable questions: "*Evidence inconclusive,* [Faith] wrote in her notebook … *Observation unreliable. Objectivity compromised*" (p. 401, original emphasis).

Sophie Anderson's *The Girl Who Speaks Bear*: A Split Vegetal Personality

The plot of Anderson's *The Girl Who Speaks Bear* (2019) also revolves around an agentic tree, yet one that is virtually diametrically opposed to Hardinge's Lie Tree. Anderson's Lime Tree is equally situated in a fairy tale world, but clearly on the "good" side. Anderson's novel is constructed as a sequel to a Russian folk tale, "The Lime Tree",[2] which features a magic lime tree whose ability to grant wishes is exploited by a greedy peasant. Finally, the exasperated tree turns the peasant and his family into bears. Anderson takes up the tale two generations later, relating the story of Yanka, the fairy tale protagonists' granddaughter and human-bear hybrid, who sets out to break the family curse. Yet after having met the talking tee, Yanka comes to realize that the alleged curse might actually be a gift.

Anderson's Lime Tree stands out for its contradictive dual nature that both enables and severely compromises its agentic potential. The lime tree in the original fairy tale blends the images of the "wish tree" and the Tree of

Life (Cooper, 2013, p. 319), but the actual plot focuses almost exclusively on the former, so that the two dimensions do not come into conflict. The tree's agency is hence limited to a single, thoroughly anthropocentric dimension: to interact with humans, and ultimately to teach them moral lessons. Anderson, in contrast, retains the wish tree dimension, but puts a much stronger stress on the tree's dimension as a Tree of Life. As a wish tree, the tree is virtually omnipotent: it can grant health and riches, and even transform humans into bears. As a Tree of Life, the "soul of the forest" (Anderson, 2019, p. 75), the Lime Tree both fulfils a different function and shows a different agentic potential. Centrally, it upholds the ecological balance of the surrounding natural system. When the angered peasant chops of one of its branches this act has catastrophic consequences for the entire ecosystem the Lime Tree symbolizes. In the chain of events started by the peasant's act of vandalism the tree is paradoxically both central agent and pointedly powerless. Incapable of controlling the fire dragon guardian that sprouts from the severed branch, the tree helplessly witnesses how this animal rather than vegetal plantlet unleashes a destructive potential that runs counter to its mother plant's wishes and in consequence the needs of the ecosystem. The tree's only remaining agency is in its voice: it can plead for Yanka's help. In every other respect it is thoroughly dependent on the human girl. For most of its appearances in the story, the lime tree is a helpless piece of wood, pointedly even described as looking "like a huge cut log" (p. 366) that needs to be dragged around by the human and animal agents of the story. The narrative thus virtually endows the Lime Tree with a split personality: on the one hand it is a powerful and agentic, but anthropocentric "wish tree", on the other an ecocentric, but simultaneously powerless and non-agentic Tree of Life.

The story evidently constructs the anthropocentric "wish tree" as a narrative to leave behind in favour of the ecocentric Tree of Life. Most of the tree's wish tree activities are situated in the mythological past, framed in interjected stories-inside-the-story that all begin in proper fairy tale mode, with "once upon a time" (p. 75). Nonetheless, Yanka, like all other human protagonists, is at first pathologically fixated on the tree's "wish tree" dimension. Part of Yanka's vital learning process is to acknowledge that the tree is not primarily a medium to fulfil her desires, but the personification of her "home" (p. 203), the Forest, which needs her agentic intervention to avert ecological catastrophe. Parallels to human exploitations of nature and to the need to re-think the human–nature relationship in the face of climate change are certainly not coincidental. Once Yanka has learned this didactic lesson, and in the process has come to realize that she can blend human and animal identity, the Lime Tree even decides that it wants to grow in her garden rather than "deep in the forest" (p. 402). This reinforces the story's ecocentric message, outlining that there is no dichotomical nature/culture border cutting of the human settlement and its inhabitants from the forest surrounding it. Rather than becoming a synanthropic plant, the Lime Tree realigns the village with the natural space surrounding it. Crucially, this is the moment when the tree's "wish tree" dimension entirely evaporates. Once the Lime Tree has taken root in the garden, the narrative does not mention its ability to grant wishes again, as if this power never existed or has self-evidently evaporated. Chronologically speaking, the tree hence develops from an anthropocentric wish tree in the fairy tale stories-inside-the-story into, at the resolution of the novel, an ecocentric Tree of Life that has shed even the last remnants of the original narrative attached to it.

Yet, concomitant with this shift in vegetal identity goes a shift from an agentic towards a non-agentic plant. On the one hand, this pointedly stresses the need for human action to avert natural disaster. Humans have brought about the catastrophe that literally sets the protagonists' world on fire, and it is them who need to restore natural balance by taking responsibility for the tree. But at the same time this re-framing of the Lime Tree's role denies the vegetal world, or, in metaphorical alignment, the ecosphere the tree stands representative for, any agency independent from humans. Even the animal protagonists, albeit far more agentic than the tree, are ultimately dependent on Yanka, and the most central didactic message of the book is that Yanka has to accept both her hybrid identity and the fact that she needs her friends and family to succeed in her endeavours. In its focus on the heroine's self-realization, the story's ideological framework reaffirms one anthropocentric narrative even whilst discarding another one. In the end, the story is always Yanka's. The tree's agency is sacrificed so that hers can flower.

Disentangling Plant Agency in Twenty-First Century Children's Literature

Van Cleve's, Hardinge's and Anderson's children's fictions are connected by a shared focus on vegetal individuals and their agentic capacities and thus make a contribution to counteract what Laist somewhat poetically refers to as the "defoliation of the cultural imagination" (Laist, 2013, p. 10). Although the three children's stories make up just three tiny samples of a vast and continually growing field of fantastic children's literature, the attention these stories pay to plant characters shows that there are at least currents in contemporary children's literature that seek to acknowledge vegetal agency and grant plants their place as, in Donna Haraway's words, "consummate communicators in a vast terran array of modalities, making and exchanging meaning among and between an astonishing galaxy of associates across the taxa of living beings" (Haraway, 2016, p. 122). In some ways, these children's stories hence challenge and subvert traditional readings of plants in literature and suggest new ways of making sense of human–plant ties. All three stories elevate plants from the underrated role of "setting" into the more prestigious role of character and give them an agency that helps shape the respective plot. In doing so they render plants more visible, seeking to raise their audience's awareness of their own plant blindness.[3] The stories also acknowledge the relevance of plants: the vegetal individuals actively shape the environmental systems they participate in.

Yet, all three texts also severely limit the parameters of vegetal agency. Ultimately, the stories remain firmly anthropocentric regarding the question whose story is actually told, and rooted in a thoroughly humanist world view. From a narratological point of view, it is quite telling that in none of the three stories a plant focalizes. They might narrate—Anderson's Lime Tree tells Yanka its history, for example—but the reader is not invited to perceive the world from a vegetal perspective. The plants are hence allowed to act, even to tell, but they are not granted a point of view. Focalization remains firmly anchored in the human sphere. The same is true for the stories' didactic messages. All three stories build on strong didactic undercurrents, and the message is always the same: a girl has to uncover her true potential. She does so by finding herself in the centre of a network of which the agentic plant is one part, but the focus of attention is always primarily directed towards her, not the plant. In all three children's books, humans remain firmly in focus. Thus, even whilst giving a voice to plants, the stories ultimately revert back to translating the vegetal individual into a metaphorical extension of the human Self.

The stories also share their project of creating a fairy-tale-like structure that frames—and also in some ways entraps—the agentic vegetal individual. The fairy tale framing is most obvious in the fairy tale adaptation *The Girl Who Speaks Bear*, but also structures the other two stories. In *The Lie Tree* the Christian mythology of the Tree is dependent on for its survival is equated with a fairy tale, in *Drizzle* the farm itself is a magic oddity disrupting the otherwise realistic world of the story. The closer one gets to the vegetal agent, the more fairy tale elements enter the respective story. This renders vegetal agency extraordinary even within the story-worlds, as supernatural, or even unnatural. The link between vegetal agency and the unnatural is overwhelming in *The Lie Tree*, as the tree is the agent who facilitates the protagonist's father in fabricating fake evidence to counteract the evolutionary theory and hence holds back the world from its "natural" evolution towards recognizing the truth in the evolutionary theory. Similarly, though less dramatically, the rhubarbs in *Drizzle* are part of an unnaturally static system that runs counter to the region's "real" climate and can only be uphold by human magic, and the Lime Tree in *The Girl Who Speaks Bear* can only retain its agency once it acts as a wish tree. As the Tree of Life—the function the story presents as the tree's "natural" one—all agency evaporates.

Ultimately all three stories spiral back towards traditional readings, not only of the vegetal world, but of ecosystems as such. Van Cleve, Hardinge and Anderson either tie agentic plants into a concept of nature as "static", or present the agency they grant their vegetal characters as "unnatural" and hence problematic. Hardinge's story sticks out from the other two in that those two stories actually appear to believe in such a static system underlying even their most agentic plant protagonists, and present this romanticized vision of nature as reality. *The Lie Tree*, in contrast, presents an ever-changing, profoundly unstable natural system, though in its equation of adaptation and progress the book also adheres to a world view that ecology as a science has moved beyond. But, crucially, *The Lie Tree* does not present agentic plants as a part of this progressive system. To the contrary, they are an essential aspect of the very regime of truth that opposes its recognition. In all three narratives, agentic plant characters hence remain tied into static frameworks, they may strain against, but ultimately fail to break free from. Vegetal agency hence extends some creepers into twenty-first century children's literature, but—at least in this very limited corpus of three books—as yet rather tentatively. Vegetal individuals remain firmly subordinate to both genre conventions of children's self-realization narratives, and to conventional myths of ecological stasis.

Notes

1 The close conceptual link between changeability/evolution and progress is a rather disturbing undercurrent in the book. In the world of *The Lie Tree*, evolution necessarily means that "everything [*is*] getting better, inch by inch, so slowly that she [can] not see it" (p. 409, original emphasis). Apparently the text confuses adaptation with progress.

2 For a summary of the complete tale see Cooper, 2013, p. 318.

3 Albeit not without falling victim to it themselves at times: Anderson's Yanka, for example, when first facing the magical tree she and her talking animal companions after all set out to ask a question all along, "stare[s] at it in wonder" (p. 344), completely flabbergasted by the fact that "it's talking" (p. 344).

References

Anderson, S. (2019). *The girl who speaks bear*. London: Usborne.

Clark, J., and I. E. Castro. (2019). Introduction. Zuzu's birds and Scout's mockingbirds: The legacy of children's agency in popular culture. In I. E. Castro and J. Clark (eds.), *Representing agency in popular culture* (pp. xi–xxxi). Lanham, MD: Lexington Books.

Cooper, P. G. (2013). The lime tree. In *Critical survey of mythology and folklore: World mythology – central and eastern Europe* (pp. 318–319). Pasadena, CA: Salem Press.

Cronon, W. (1996). Introduction: In search of nature. In *Uncommon ground: Rethinking the human place in nature* (pp. 23–56). London: Norton.

Foucault, M. (2002). Truth and power. In J. D. Faubion (Ed.), *Essential works of Foucault 1954–1984*, vol. 3: *Power* (pp. 111–133). Harmondsworth: Penguin.

Garrard, G. (2012). *Ecocriticism*. London: Routledge.

Haraway, D. (2016). *Staying with the trouble: Making kin in the Chthulucene*. Durham, NC and London: Duke University Press.

Hardinge, F. (2015). *The lie tree*. London: Macmillan.

Hogan, W. (2009). *Animals in young adult fiction*. Lanham, MD: Scarecrow Press.

Höing, A. (2017). *Reading divine nature*. Trier: WTV.

Johnson, K. R. (2000). *Understanding children's animal stories.* Lewiston, NY: Edwin Mellen Press.

Keetley, D. (2016). Introduction: Six theses on plant horror; or, why are plants horrifying? In D. Keetley, A. Tenga (Eds.), *Plant horror: Approaches to the monstrous vegetal in fiction and film* (pp. 1–30). London: Palgrave Macmillan.

Laist, R. (2013). Introduction. In R. Laist (Ed), *Plants and literature: Essays in critical plant studies* (pp. 9–18). Amsterdam and New York: Rodopi.

Plumwood, V. (1993). *Feminism and the mastery of nature.* London and New York: Routledge.

Ryan, J. C. (2018). *Plants in contemporary poetry: Ecocriticism and the botanical imagination.* New York: Routledge.

Soper, K. (2015). The discourses of nature. In K. Hiltner (Ed), *Ecocriticism: The essential reader* (pp. 267–277). London: Routledge.

Van Cleve, K. (2010). *Drizzle.* New York: Puffin.

12 Vegetable Violence

The Agency, Personhood, and Rhetorical Role of Vegetables in Andy Griffiths and Terry Denton's *The 52-Storey Treehouse*

Lykke Guanio-Uluru

With their bestselling illustrated book series about a multi-storied treehouse, Australians Andy Griffiths (author) and Terry Denton (illustrator) have reached an international audience. The books have been criticized for their use of subversive and scatological humour, their irreverence for authority figures, and their perceived encouragement of young readers toward bad behaviour (as summarized in Hately, 2016). This article focuses on the fourth book in the series of (currently) nine books, *The 52-Storey Treehouse*, which features multiple episodes of violence by and against vegetables and has a plot line that sends the series' protagonists travelling to the Vegetable Kingdom. With reference to the field of critical plant studies (Hall, 2011; 2019; Manusco & Viola, 2016; Marder, 2013; Nealon, 2016; and Ryan, 2016; 2018), and drawing on Kerry Mallan's (1993) discussion of humour in children's literature, this article analyses the attribution of personhood and agency to vegetables in the story, examining their rhetorical role in Griffiths and Denton's book, with an emphasis on acts of vegetable violence. The analysis demonstrates that *The 52-Storey Treehouse*, through the attribution of personhood to vegetables, leads the reader on a multifaceted exploration of their moral standing.

While quite a few studies discuss the role of food in children's literature (see for instance Keeling & Pollard, 2009;[1]Carrington & Harding, 2014), few contributions yet involve other-than-human perspectives on human relationships to food. There also seems to be little done on the representation of vegetables,[2] which is interesting, given that the dinner battle over vegetables is well known to most parents. While a discussion of plant agency and personhood in *The 52-Storey Treehouse* in relation to tropes of food in children's literature is outside the scope of this chapter, its analysis founded in critical plant studies may introduce new kinds of "eat-your-vegetables" dinner table discussions, as foreshadowed by the treatment of vegetables in Griffiths and Denton's narrative.

Critical Plant Studies, Plant Agency, and the Concept of Vegetable Violence

In the fourth book in the wildly popular Treehouse series, *The 52-Storey Treehouse*, the series' protagonists Andy and Terry go in search for a prince

DOI: 10.4324/9781032066356-17

to awaken their sleeping neighbour, Jill. Attempting to outsmart an ugly potato prince in the Vegetable Kingdom, they soon find themselves trapped by vengeful vegetables. Casting plants in the role of villains, the story undermines the positioning of plants as helpless victims, pointing to their considerable resources and their independent agendas. The protagonists' rescue comes in the form of the violent Vegetable Patty, who turns out to hate vegetables even more than Andy and Terry do. In fact, Patty enacts such a wide register of vegetable violence that the series' protagonists begin to sympathize with the cultivated produce. Are Griffiths and Denton standing up for plants, or are they, on the contrary, making fun of ideas proposed by the rapidly expanding field of critical plant studies?

While there have long been voices proposing that plants are sentient and possess "some degree of voluntary powers" (Darwin, E., 1796), expressed not least through root systems functioning akin to a brain (Darwin, C., 1880), such suggestions have not, so far, gained traction with the scientific mainstream. Recently, however, biological research has emerged to (re)demonstrate the variety and complexity of the adaptive behaviour of plants, leading to claims of plant intelligence (Manusco & Viola, 2016; Pollan, 2013; Trewavas, 2003) as well as to a more pervasive botanical and philosophical questioning of the predominant Western cultural attitude towards, and representation of, plants.

This upsurge of interest in plants is understandable as the cumulative effects of ecocriticism (see for instance Carson, 1962; Næss, 1989; Buell, 1995; 2005, and Glotfelty and Fromm, 1996), animal activism (Singer, 1975; Derrida, 2002; Haraway, 2008), and critical and feminist posthumanism (Haraway, 1991; Hayles, 1999; Wolfe, 2010; Braidotti, 2013), which have gained momentum as persistent critiques of the modern capitalist exploitation of both the environment and other species. The research into plants and their capacities has gained additional urgency in the context of the global climate change crises, since "plants dominate every terrestrial environment, composing ninety-nine percent of the biomass on Earth" (Manusco and Viola, 2016, p. xii) and thus are vital to the global climate. Can plants save us from the consequences of our environmental follies? Their biosphere dominance at least suggests a natural resilience that seems hopeful in this age of accelerated species loss.

Biosphere plant predominance notwithstanding, are we free to behave towards plants in the manner that best suits our needs? If animals are now being included in the moral spectrum, following the battle by animal activists, plants seem the next logical extension of a species-inclusive posthumanist sensibility—a point also made by Michael Marder in the introduction to *Plant Thinking: A Philosophy of Vegetal Life* (2013, p. 2) and suggested by Donna Haraway's call for the making of kin across species (2016). The position is not uncontroversial, however. Discussing the pervasive sidestepping of the question of plant life in Western philosophy, and with reference to Michel Foucault's concept of biopower, Jeffrey T. Nealon notes that:

> ... animal studies' blanket refusal to consider vegetable life within its biopolitical frame seems to function as a subset of an old practice: trying to close the barn door of ethical consideration right after your chosen group has gotten out of the cold of historical neglect.
>
> (Nealon, 2016, p. xii)

The conflict of plant versus animal rights is thematized in *The 52-Storey Treehouse* through the plot line that sees Andy and Terry's animal loving neighbour Jill, who lives with a menagerie of cats, rabbits, dogs, horses, a goat, and a cow, put to sleep as she pricks herself on a pointed carrot, Sleeping Beauty-style. Her prolonged slumber is eventually discovered by Andy and Terry, as they, dressed in pith helmets, hack their way into her overgrown house with a set of machetes. While their attire is a visual nod to jungle and colonist narratives, they are also inflicting damage on plant life. Later, when the three of them find themselves trapped and imprisoned in the Vegetable Kingdom, and stand accused of several acts of hatred towards and violence against vegetables, Jill pleads to be freed since she would "never *dream* of hurting" vegetables (p. 220). From earlier books in the series, the reader knows that Jill can talk to all kinds of animals and insects, such as sharks and ants—a purpose for which she is sometimes summoned to the treehouse by Andy and Terry. However, it now becomes clear that her kindness does not extend to vegetables, so her plea to be released falls on deaf ears: "No Way! ... You feed truckloads of poor defenseless vegetables to your animals every day. That's why we sent that cursed carrot to your house to put you to sleep" (p. 220).

The story line of course contradicts the claim that the vegetables are "defenseless" by revealing that their agency extends well beyond their own Kingdom. The vegetables' ingenuity in carrying out their scheme paints them as possessors of plant intelligence, and their vengeful behaviour contradicts a culturally predominant view of plants as passive, silent and pleasing (Ryan, 2018, p. 1). Still, the vegetables' accusation towards the otherwise considerate Jill calls to mind the complaints of maltreatment on the part of the Old Forest in J. R. R. Tolkien's *The Lord of the Rings* (1954–1955) and echoes Marder's argument that plants have been subject to ethical neglect and exploitation:

> In the absence of the will to think through the logic of vegetal life, beyond its bio-chemical, cellular or micro-molecular processes and ecological patterns, philosophers readily assumed that within the broad evolutionary frame of reference, the existence of plants is less developed or less differentiated than that of their animal and human counterparts and that therefore vegetal beings are unconditionally available for unlimited use and exploitation. Such repression of the most basic question regarding plants became the breeding ground for their ethical neglect ...
>
> (Marder, 2013, pp. 2–3)

Throughout the Treehouse series, Jill is depicted as ethically considerate towards animal life. She too, however, in caring for her animal friends, has failed to consider the moral standing of plants and this shortcoming is, humorously but somewhat edgily, thematized in *The 52-Storey Treehouse*.

Certain aspects of plant representation in the Treehouse series suggest traces of plant blindness (Wandersee & Schussler, 1999; Wandersee & Clary, 2006; White, 2019). While Western literature has been more attentive to the plant world than Western science and philosophy, it still tends to conform to a general anthropocentric paradigm, as literary discourse predominantly makes symbolic or figurative use of plants to represent human and animal meaning, treating the landscape as the backdrop for human or animal dramas (Gagliano et al., 2017, pp. x–xi) and this form of backgrounding is the fate of the tree sustaining Andy and Terry's sprawling treehouse. The series' titles, as well as the early presentation of the treehouse tree at the start of each book in the series, right after the habitual presentation of Andy and Terry, seem to suggest that the tree is nearly as central to the story as its protagonists. However, it is the treehouse as a human construction, rather than the tree itself that receives the protagonists', and therefore the readers' attention: "Well, when I say 'tree', I mean treehouse" (p. 4).

The front endpapers of each book in the series show a cross-section of the crown of the tree, to display the increasingly extravagant contents of the treehouse, resulting from Andy's and Terry's constant expansions. With reference to John Charles Ryan's distinction between the *extrinsic* (forces exerted on plants) and *intrinsic* (attributes generated by plants themselves) capacities of plants (2016, p. 41), the treehouse remains acted upon by Andy and Terry throughout the series, while they depend, without acknowledging this fact, upon the intrinsic ability of trees to grow upward and expand. The overall role of the tree in the series is thus to serve as "the backdrop for human or animal dramas", consonant with the anthropocentric perspective common to Western literature and culture.

Although trees have been singled out by many cultures as worthy of particular attention, as testified by the symbolic significance they have in many religious traditions (Guanio-Uluru, 2015; Guanio-Uluru, 2018), the tree is but a silent background figure in the Treehouse series. Zoe Jaques notes that trees can "be said to have the most symbiotic of relationships with humans, and they also generate as much affection as pets, thereby deconstructing a humanistic hierarchy of being which traditionally places fauna above flora" (2015, p. 113). Drawing on Maud Hines (2004), Jaques further argues that the literary device of giving trees a voice "grants them subject status", thus making them worthy of protection (p. 113). The treatment of both the treehouse tree and the vegetables in *The 52-Storey Treehouse* runs counter to these maxims, however, since the tree has no voice or subject status in the story and the vegetables, which do appear as vocal "subjects", meet with a violent end.

Unlike the silent tree, the vegetables portrayed in prose and imagery in *The 52-Storey Treehouse* are neither innocent nor defenceless and vegetable violence, as the concept is explored here, is thus a two-way street: it is the violence performed both by and against vegetables in the story. While pitting the vegetables in the narrative as agential beings, *The 52-Storey Treehouse* acknowledges, not least through the character of Vegetable Patty, that vegetables usually are acted upon by humans, and frequently so in a violent way. However, as Vegetable Patty fights to eradicate vegetables from the face of the earth (p. 248), the vegetables in *The 52-Storey Treehouse* fight back, exerting their forces on the story's protagonists.

Vegetables as Persons: Giving Plants a Voice—and a Face

Within the field of critical plant studies, there is ongoing discussion concerning whether the moral standing of plants can be legitimized by thinking of plants as persons, an argument furthered by Matthew Hall (2011), who has examined attitudes to plants in Eastern, Western, Indigenous and Pagan systems of thought, and holds that plants are other-than-human persons that should be recognized as relational, intelligent beings worthy of our respect. Countering Hall, Marder has proposed that the attribution of personhood to plants is a form of metaphysical violence, "forcing plants into the mold of appropriative subjectivity" (2013, p. 55) by imposing human philosophical categories upon them. Hall has refuted Marder's criticism by highlighting that his discussion of plant personhood has an animist orientation, rather than a Western philosophical basis, and that it is inspired by indigenous cultures that do not equate "person" with "human" but bases the notion of personhood on kinship relationships (2019). While Western philosophical notions of personhood have centred on the attribution of certain mental capacities, privileging reason and self-consciousness, Hall discusses by example several animist traditions where "The category of person applies to anything that has being, and who is capable of relating" (Detwiler, 1992, p. 239, quoted in Hall, 2019, p. 5), emphasizing that "a person is one who is in a kinship relationship, not necessarily based on genealogy, mythology or biology, but through the *situated act* of sharing and relating" (Hall, 2019, p. 5, my emphasis). These distinctive positions in regard to plant personhood help clarify the ideological underpinnings of the episodes of vegetable violence in *The 52-Storey Treehouse*.

Along with much children's literature, *The 52-Storey Treehouse* resorts to a degree of anthropomorphism as a storytelling strategy in the villainization of, and call for empathy with, vegetables. Anthropomorphism is here understood as "the interpretation of nonhuman things or events in terms of human characteristics" (Guthrie, 2008). Since *The 52-Storey Treehouse* is a visual children's novel (Tønnesen, 2018, p. 181), characterized by presenting a single, longer, unified narrative published in a book-format, where script, images and design "are so closely knit together that they only make sense when read as a visual unity" (Tønnesen, 2018, p. 173, my translation), this anthropomorphism is developed both visually and verbally. The analysis of the narrative's position(s) on plant personhood and vegetable violence thus requires a detour explaining certain characteristics of visual children's novels, including a brief discussion of the story's humorous tone.

Jeff Kinny's series *Diary of a Wimpy Kid* and Liz Pichon's *Tom Gates* series are well-known examples of visual children's novels. Kai Mikkonen further cites Kinny's series as an example of a hybrid form of the comic, describing the series as "books illustrated with comics and cartoons" (2017, p. 13), a characteristic that applies equally well to the Treehouse series. Most of the illustrations in *The 52-Storey Treehouse* anchor and expand on the narrative presented in the verbal text. At intervals, however, the visuals are whole page spreads, carrying the main weight of the storytelling. Such page spreads invite what Mikkonen terms a global reading, that is, a nonlinear way of reading, as opposed to the sequential reading that is common in comics (2017, p. 36). The sequential sections in *The 52-Storey Treehouse* include the excerpts from Vegetable Patty's book *Fun with Vegetables* (pp. 85–103), Andy's and Terry's journey through the Vegetable Kingdom (pp. 172–181), and the section that shows Patty demolishing vegetables within the Vegetable Kingdom (pp. 234–239). Consequently, the parts of the book that most explicitly deal with violence by and towards vegetables are presented in cartoon panelling, inviting a sequential reading, with an emphasis on movement, action, and repetition.

The reading of comics involves attending to the relationship between panels and to the juxtapositions and transitions between these. The analysis of verbal–visual relationships involves inferences about changes in viewpoint through careful attention to visual cues, such as facial features, gestures, and other details (Mikkonen, 2017, p. 40). In this reading, the vegetables' facial expressions and their body language are of particular importance to the interpretation of the narrative's overall rhetoric concerning the narrated acts of vegetable violence in *The 52-Storey Treehouse*. Thus, when discussing the illustrations, this analysis emphasizes the focalized (Bal, 2006), in accordance with Mikkonen's argument that in graphic storytelling, "the image necessarily reveals a spatial point from which something (the focalized) is perceived" (Mikkonen, 2017, p. 74), so that the focalized matters more to the narrative cohesion than the focalizer. In hybrid comic forms such as *The 52-Storey Treehouse*, Hallberg's concept of iconotext (1982), that is, the conception of text and image as a unity of meaning, complements the vocabulary of comic narrative theory as developed by Mikkonen.

In *The 52-Storey Treehouse*, the theme of vegetable violence is first introduced in an illustration of four small animals that are busy loading a carrot gun. A tiny caption reads: "spare carrots" (p. 7), with an arrow indicating a stack of carrots nearby. The violence soon develops to include the human species, when a couple of pages later, Andy and Terry are displayed juggling chainsaws. Two speech bubbles establish a level of intra-diegetic commentary as the visual character Andy screams "Terry! I just chopped off my nose!" The visual character Terry replies, in another speech bubble: "I can't hear you, Andy. I just chopped off my ears!" (p. 9). Such episodes establish casual violence, characteristic of slapstick comedy, as an integral part of the story world. According to Kerry Mallan, such slapstick or banana-skin humour "provides the basic ingredient of much humorous literature for children" (1993, p. 34). For smaller children, the illustrated format is of importance as their appreciation of this type of humour "is often triggered by the visual realisation of what's happening; the pictures are the stimulus for their laughter" (Mallan, 1993, p. 34). Mallan adds that older children probably handle such depictions better, since they are able to distinguish between reality and fantasy and thus can appreciate parody and gross exaggeration (1993, p. 7), such as the violent absurdities depicted in the scene just discussed. The vegetable violence within the story consequently occurs within a context where cartoon violence is the norm rather than the exception and where the illustrations often provide the exaggeration that turns the verbal narration into slapstick comedy. In the cited instance where carrots are used as ballistic projectiles, the carrots are not attributed agency, nor do they display any anthropomorphizing characteristics.

The first sign of vegetable personhood in the book coincides with the reader's first encounter with the narrative's most aggressive perpetrator of vegetable violence. This incident occurs when Andy and Terry visit the office of their publisher, Mr Big Nose, to discover why he has not called as per usual to remind them of the imminent deadline for their manuscript.[3] The illustrations show his office to be uncharacteristically untidy and among scattered objects and quite a few leaves on the floor they discover the book *Fun with Vegetables*, by Vegetable Patty. Its front cover depicts an angry woman, dressed in armour and brandishing savage-looking medieval weapons. Three, open-mouthed, scared-looking carrots "run" with beads of sweat around their "heads" in flight from the vengeful, sneering Patty (p. 86). The reader gets to read the start of the book along with Andy and Terry and the book-within-the-book consequently hands narration over to Patty: "Hi, Vegetable Patty here. As we all know, vegetable fighting is a serious business, but that doesn't mean it can't be fun. Don't believe me? Well, read my book!" (p. 87). The change of narrator from Andy to Patty suggests that the anthropomorphized vegetables are focalized through Patty and may be attributed to her personal views. However, when Andy and Terry later enter the Vegetable Kingdom, they too encounter personified vegetables "in the flesh". All the same, the rhetorical choice of inventing a special character responsible for gross acts of vegetable violence indicates a distancing mechanism by which Andy and Terry need not be complicit in such acts.

The excerpts from Patty's book cover 17 pages and display Patty as she demolishes vegetables in multiple ways. This part of *The 52-Storey Treehouse* is framed in cartoon gutters. The verbal narration is a brief but rhythmically satisfying series of "war-calls" that rhyme:

> "Boil them! Broil them! Salt and oil them! Crunch them! Munch them! Knock-out punch them! Grab them! Stab them! Shish kebab them! Throw them! Mow them! Tae-kwon-do them! Kick them! Flick them! Pogo stick them! Mash them! Smash them! Whip and lash them! Whack them! Smack them! Sap lock sac them! Crush them! Mush them! Drown and flush them!"
>
> (pp. 88–103)

Each set of three imperatives cover a doublespread, with evocative illustrations of vegetables being acted upon in the specified way by Patty, who effectively reduces the anthropomorphized vegetables to vegetable patties (suggesting perhaps, that Patty engages not just in physical violence but also in a "metaphysical violence" as defined by Marder (2013, p. 55), by making the vegetables more appellatively like herself).

Only 5 of the 24 accompanying illustrations display the vegetables with no sign of personalization. In most of the cartoon drawings, the vegetables are depicted with anguished and frightened facial expressions, thus invoking sympathy, at least in a sensitive reader. Terry embodies this reading position: "'That's enough, Andy,' says Terry, 'I can't take any more. It's too violent! I never thought I'd say this, but I actually feel sorry for those poor vegetables …'" (p. 104). The illustration shows Terry covering his eyes with his hands. Even Andy agrees: "'Me too,' I say. 'It's kind of weird to feel sorry for something you hate so much'" (p. 105).

It is the attribution of personhood on the part of the vegetables, visible in their anguished faces, that form the basis for this sympathy. This suggests that the personhood attributed to the vegetables is founded on a recognition of their cognitive capacities, in this case the ability to feel fear and pain—a position compatible with Western philosophical notions of personhood. The sympathy develops in response to a literary representation of vegetable pain and thus becomes a metacommentary to the treatment of vegetables in *The 52-Storey Treehouse*. Hall emphasizes "situated acts of sharing and relating" (2019, p. 6) as constitutive of (cross-species) kinship, and Andy's and Terry's responses to the (representation of) suffering vegetables indicate the beginnings of a relationship to vegetables that is becoming sensitized to their potential personhood.

The protagonists' reactions to Patty's vegetable brutalities appear as an unusual squeamishness to violence on their part. In her discussion of humour in children's literature, Mallan finds that there are limits "to the nature and extent of the violence that will be tolerated, but situations which allow characters to emerge relatively unscathed are treated light-heartedly" (1993, p. 7). Such inability of violence to cause lasting harm is the norm in the rest of the narrative: While in the chainsaw episode Andy is depicted with a chopped off nose and Terry with a hole in his head where his ear has been, they bear no traces of the violence on the subsequent doublespread and it is consequently exposed within the fiction as an absurdity. From a rhetorical perspective, the "Patty-cartoon" thus constitutes a break with the slapstick comedy that characterizes the rest of the book in two ways. Firstly, Patty's violence against vegetables induces sympathy. Generally, Andy and Terry do not feel sorry for each other (or others) when they are subject to violence, such as in the incident when they chop off their own nose and ears. Secondly, the (same) vegetables do not reappear unharmed in the next spread. This indicates that Patty's vegetable violence is of a different kind: it is *real* violence, in that its effects are both felt and lasting. The depicted violence is also more real in that, with a few creative exceptions, these are types of violence that vegetables undergo every day. Here, the vegetables are framed as victims—hence the need for the protagonists' pity.

Walking a Mile in their Moccasins: Phytomorphism in the Vegetable Kingdom

The vegetables' victimized position is overturned as Andy and Terry enter the Vegetable Kingdom, where the vegetables are revealed as powerful agents. In fact, they turn out to have kidnapped Mr Big Nose, further suggesting a power that infiltrates and impacts the human world (albeit at first glance in a manner more consistent with detective fiction than with botanically accurate plant agency). Andy and Terry, with Jill in tow in a glass coffin, make their entry into the Vegetable Kingdom on the trail of a (very hungry) caterpillar, a nod to Eric Carle's 1969 classic. Walled in by a giant field of asparagus, the Vegetable Kingdom is closed off, like Jill's overgrown house after she has fallen prey to the cursed carrot. The pattern repetition suggests that thick walls of plants are a common plant defence strategy. The asparagus, like the plants covering Jill's house, is depicted without anthropomorphizing features. This changes as Andy and Terry break through the wall of asparagus to discover "a garden full of vegetables!" (p. 180). While there is nothing unusual in this verbal announcement, the illustration displays a more imaginative scene of smiling vegetables, engaged in the quintessentially British pursuits of picnicking, and playing tennis and croquet. The illustration thus confers both agency and personhood on the vegetables, anthropomorphizing them by showing them engaged in "civilized" human activities.

To blend in, the human protagonists decide to disguise as vegetables: Terry dons a broccoli costume, Andy takes on the form of a cob of corn and the sleeping Jill is dressed up as a carrot. Consequently, while the vegetables are anthropomorphized in the illustrations, the story's main protagonists are "phytomorphized", suggesting a two-way influence between the human and vegetable kingdoms. Through "species cross-dressing" the human protagonists take on the perspective of vegetables, even if the identification is only "skin deep". While they change "skins", they do not have to change their *behaviour* to fit in, since the vegetables already behave like humans. Arguably, this moves the relationship between humans and vegetables in the story closer to a sense of what Hall identifies as "we-ness": "A focus on interrelation rather than on difference [that is] an integral part of animist conceptions of personhood" (2019, p. 7).

Initially, the scheme is successful, as the Kingdom's ugly potato prince soon wishes to marry carrot-Jill, who thus gets her Sleeping Beauty kiss and wakes up. However, when Jill pulls off her costume to prove she is not a vegetable, the scam is revealed. As a result, they are all seized by a "brutal battalion of brussel sprouts" (p. 192) and thrown in the dungeons, where they find Mr Big Nose, who has been kidnapped for publishing Patty's *Fun with Vegetables*. In a sudden turn-around, the initially victimized vegetables are now cast in the role of aggressors, charging the human intruders with vegetable-impersonation (p. 194). Soon, they face "a mob of angry vegetables" (p. 215), and, in a role reversal, are placed into a pot to make human soup—a visual allusion to what Martine Hennard Dutheil calls the "explorer's fear of ending up simmering in a giant cooking pot, an all-time favourite cliché of popular colonial fiction" (2001, pp. 105–106). By playing repeatedly on such colonialist clichés, the illustrations develop a parallel interpretive frame, casting Andy and Terry as colonizers and explorers of the Vegetable Kingdom, thus alluding to their ethically problematic position in relation to the vegetables.

From the perspective of critical plant studies, the display of vegetable aggression mirrors facets of actual plant behaviour. At least since the publication of Charles Darwin's *Insectivorous Plants* (1875) it has been known that certain species of plants lure, trap and devour insects and that there are carnivorous plants that even trap and digest small animals (Manusco & Viola, 2016, p. 62). Several illustrations at this stage show the vegetables with toothful grins, aligning them with the trope of the monster plant, which, as T. S. Miller (2012) has suggested, plays on human taxonomic unease vis-à-vis plants. One of the leering plants is reading a book with the modified title: "fun with ~~vegetables~~ humans" (p. 215). The carnivalesque role reversal is further developed as the vegetables dance around the pot and sing: "We know just how you feel: You don't want to be a meal! You don't want your skin to peel! You don't want to scream and squeal!" (p. 224). While this may be gloating, it also suggests vegetable sympathy with their human victims, reflecting Terry's and Andy's earlier sympathy with the vegetables molested by Patty. In light of the narrative progression, the chant "we know just how you feel" further strengthens the sense of "we-ness" between humans and vegetables and the whole sequence, with Andy and Terry dressing up as vegetables, the forceful agency of the vegetables themselves, and the role reversal in the soup scene, asks the reader to take on the perspective of vegetables—a perspectival change first suggested in the expressive illustrations in Patty's book, where the vivid and agonized faces of the chopped vegetables cause Andy and Terry to pity them.

However, when all seems lost for the story's protagonists, as the soup gradually heats, Vegetable Patty comes to their rescue, slicing, dicing, whipping, and plastic-wrapping the vegetables over three doublespreads, freeing their human victims. Significantly, the representation of this final "victory" over the vegetables returns them to an inanimate, non-personal status: in these illustrations, none of the vegetables butchered by Patty are portrayed with eyes or faces, discouraging identification and sympathy with them on the reader's part, even if earlier sequences, along with the vegetables' song, invoke the brutal lot of vegetables more generally. Consequently, the narrative conclusion seems to back away from the implications of the vegetable sympathy that parts of the story seek to investigate.

Significantly, the character of Patty represents both the entry and exit point for the narrative's reflection on plant agency and personhood. Her inordinate hatred of vegetables is explained with reference to a childhood incident where both her parents were accidently squashed by a display of giant vegetables due to a freak storm—a situation where the wind, rather than the vegetables, is the active agent. Consequently, her vendetta towards vegetables is founded on a faulty logic. Patty's appearance in the narrative ends with her emphatically urging the reader to "Eat your vegetables!", in order to "wipe them from the face of the earth forever!" (p. 248). Her final message highlights the complexity of the book's address to the reader: Is it undermining the parental admonition of eating vegetables (since it comes after a plot line that encourages children to sympathize and identify with them)—or is it undermining its own attempt to examine an expanded "we-ness" that also includes vegetables and other plants? In the rest of the narrative, as the protagonists return to their treehouse, there are no further signs of plant agency or vegetable personhood—such notions are all extinguished by Vegetable Patty.

Conclusion

The positioning of the treehouse tree as a silent backdrop for the action in the rest of the series and the elimination of the idea of plants as persons at the end of *The 52-Storey Treehouse* suggests that the exploration of plant personhood is an idea that is toyed with rather than an integrated view on the part of the implied authors. Conferring agency to vegetables, the narrative explores questions regarding plant agency, plant personhood, and the moral standing of plants, moving towards a conception of plant personhood that aligns with the sense of "we-ness" of indigenous animist views (Hall, 2019). This move is accomplished through anthropomorphizing the vegetables' activities while phytomorphizing the story protagonists. However, the narrative eventually pulls back to a colonialist-like extinction of pluralism as Patty enters the Vegetable Kingdom, putting an end to the sentient life of its inhabitants. The colonial anxiety tied to the image of the soup pot thus comes to reflect unease on the part of the implied authors regarding the issue of plant agency and plant personhood. The vegetables are overall portrayed as a disruptive presence, putting the protagonists' friend Jill to sleep, kidnapping their publisher, and subjecting them all to mortal danger by threatening to devour them. Towards the end of the narrative, the animal lover Jill is rescued and restored while the vegetables are turned into patties and zip-lock bagged for human consumption. This frames the ideas of vegetable sentience, and of plants as other-than-human persons worthy of respect, as carnivalesque notions and as a temporary reversal of the status quo. While the protagonists return to their treehouse, relaxing among their assorted animal pets, plants are again relegated to the background of the story action. From the point of view of critical pant studies this ending is somewhat unsatisfactory since the exploration of plant sentience ultimately is sealed off and silenced.

While the book leads the reader through various moral positions in relation to vegetables, moving from pitiful sympathy (as co-readers of Patty's violent book) to reflections around the agency and personhood of plants and vegetables (within the Vegetable Kingdom) to an acceptance that plants may also be aggressors and "villains", the final silencing of the vegetables indicates a reluctance to accept the proposition that plants have moral worth. Even so, the vigorous agential vegetables making their guest appearance in *The 52-Storey Treehouse* manage to memorably disrupt the story universe, potentially sowing a few seeds of phyto-awareness in the reader along the way.

Notes

1 *Food in children's literature* makes not a single mention of vegetables. "Vegetarianism" occurs once, in relation to Ovid's *Metamorphosis* (p. 4).

2 *Feast or famine* has in total seven references to "vegetables", most of which are generic mentions of vegetables in conjunction with other foods. Only one of the references is relevant to this discussion: the mention of Berly Doherty's phytocentric, metaphoric use of vegetables in poetry, as in "If you were a carrot and I was a sprout, I'd boil along with you, I'd sit on your plate" (in Pinset, 2014, p. 236).

3 The frame narrative in the Treehouse series usually centres on job procrastination, as each book details the crazy adventures undertaken by Andy and Terry in the process of writing the book that the reader is currently reading, while struggling to complete and hand in the manuscript before a tight deadline.

References

Bal, M. (2006). *A Mieke Bal reader*. Chicago: The University of Chicago Press.

Braidotti, R. (2013). *The posthuman*. Cambridge: Polity Press.

Buell, L. (1995). *The environmental imagination: Thoreau, nature writing, and the formation of American Culture*. Cambridge, MA: Harvard University Press.

Buell, L. (2005). *The future of environmental criticism*. Malden, MA: Wiley-Blackwell.

Carle, E. (1969). *The very hungry caterpillar*. Cleveland, OH: World Publishing Company.

Carrington, B., & Harding, J. (2014). *Feast or famine: Food and children's literature*. Newcastle upon Tyne: Cambridge Scholars Publishing.

Carson, R. (1962). *Silent spring*. New York: Houghton Mifflin Company.

Darwin, C. (1875). *Insectivorous plants*. London: John Murray.

Darwin, C. (1880). *The power of movement in plants*. London: John Murray.

Darwin, E. (1796). *Zoonomia, Vol. I, Or, the laws of organic life*, Section XIII, "Of vegetable animation". Retrieved from www.gutenberg.org/files/15707/15707-h/15707-h.htm

Derrida, J. (2002). The animal that therefore I am (more to follow). *Critical Inquiry*, 28(2), 369–399.

Detwiler, F. (1992). "All my relatives": Persons in Oglala religion. *Religion* 22, 235–246.

Dutheil, M. H. (2001). The representation of the cannibal in Ballantyne's *The Coral Island:* Colonial anxieties in Victorian popular fiction, *College Literature* 28 (1), 105–122.

Gagliano, M. A., Ryan, J. C., & Vieira, P. (2017). *The language of plants: Science, philosophy, literature*. Minneapolis: University of Minnesota Press.

Glotfelty, C., & Fromm, H. (Eds.) (1996). *The ecocriticism reader: Landmarks in literary ecology*. London: The University of Georgia Press.

Griffiths, A., & Denton, T. (2014). *The 52-storey treehouse*. London: Macmillan Children's Books.

Guanio-Uluru, L. (2015). *Ethics and form in fantasy literature: Tolkien, Rowling and Meyer*. Basingstoke: Palgrave Macmillan.

Guanio-Uluru, L. (2018). Plant-human hybridity in the story world of Kubbe. In N. Goga, L. Guanio-Uluru, B. O. Hallås, & A. Nyrnes (Eds.), *Ecocritical perspectives on children's texts and cultures: Nordic dialogues* (pp. 125–140). Basingstoke: Palgrave Macmillan.

Guthrie, S. E. (2008). Anthropomorphism. *Encyclopaedia britannica*. Available at: https://www.britannica.com/topic/anthropomorphism

Hall, M. (2011). *Plants as persons: A philosophical botany*. Albany: State University of New York Press.

Hall, M. (2019). In defence of plant personhood. *Religions*, 10(5). doi:10.3390/rel1005031.

Hallberg, K. (1982). Litteraturvetenskapen och bilderboksforskningen. *Tidskrift för litteraturvetenskap*, 3, 163–168.

Haraway, D. (1991). *Simians, cyborgs and women: The reinvention of nature*. New York: Routledge.

Haraway, D. (2008). *When species meet*. Minneapolis: University of Minnesota Press.

Haraway, D. (2016). *Staying with the trouble: Making kin in the Chthulucene*. London: Duke University Press.

Hately, E. (2016). "'Watch your language, Andy,' said Terry. 'There might be children reading; or, Early Readers the Australian way." In J. M. Miskec & A. Wannamaker (Eds.), *The early reader in children's literature and culture: Theorizing books for beginning readers* (pp. 198–212). New York and London: Routledge.

Hayles, N. K. (1999). *How we became posthuman: Virtual bodies in cybernetics, literature and informatics*. Chicago: University of Chicago Press.

Hines, M. (2004). "He made us very much like the flowers": Human/nature in nineteenth century Anglo-American literature. In S. I. Dobrin & K. B. Kidd (Eds.), *Wild things: Children's culture and ecocriticism* (pp. 16–30). Detroit: Wayne State University Press.

Jaques, Z. (2015). *Posthumanism and children's literature: Animal, environment, cyborg*. London: Routledge.

Keeling, K. K., & Pollard, S. T. (2009). *Approaches to food in children's literature.* New York and London: Routledge.

Mallan, K. (1993). *Laugh lines. Exploring humour in children's literature.* Newtown: Primary English Teaching Association.

Mancuso, S., & Viola, A. (2016). *Brilliant green: The surprising history and science of plant intelligence.* Washington: Island Press.

Marder, M. (2013). *Plant-thinking: A philosophy of vegetal life.* New York: Columbia University Press.

Mikkonen, K. (2017). *The narratology of comic art.* New York and London: Routledge.

Miller, T. S. (2012). Lives of the monster plants: The revenge of the vegetable in the age of animal studies. *Journal of the Fantastic in the Arts*, 23(3), 460–479.

Nealon, J. (2016). *Plant theory: Biopower and vegetable life.* Stanford: Stanford University Press.

Neass, A. (1989). *Ecology, community and lifestyle.* (D. Rothenberg, Trans.). Cambridge: Cambridge University Press.

Pinset, P. (2014). A varied menu: Children's poetry about food. In B. Carrington & J. Harding (Eds.), *Feast or famine: Food and children's literature* (pp. 228–239). Newcastle upon Tyne: Cambridge Scholars Publishing.

Pollan, M. (2013). The intelligent plant. *The New Yorker*, 3 December.

Ryan, J. C. (2016). Tolkien's sonic trees and perfumed herbs: Plant intelligence in Middle-earth. In P. Vieira, M. Gagliano & J. C. Ryan (Eds.), *The green thread: Dialogues with the vegetal world.* New York and London: Lexington Books.

Ryan, J. C. (2018). *Plants in contemporary poetry: Ecocriticism and the botanical imagination.* London and New York: Routledge.

Singer, P. (1975/2002). *Animal rights: A new ethics for our treatment of animals.* New York: Harper Collins.

Tolkien. J. R. R. (2005/1954–55). *The lord of the rings.* (50th anniversary ed.). London: HarperCollins Publishers.

Tønnesen, E. S. (2018). Visuelle barneromaner. In R. S. Stokke & E. S. Tønnesen (Eds.), *Møter med barnelitteratur: Introduksjon for lærere* (pp. 191–194). Oslo: Universitetsforlaget.

Trewavas, A. (2003). Aspects of Plant Intelligence. *Annals of Botany*, 92, 1–20.

Wandersee, J. H., & Schussler, E. E. (1999). Preventing plant blindness. *The American Biology Teacher*, 61(2), 84–86.

Wandersee, J. H., & Clary, R. E. (2006). Advances in research toward a theory of plant blindness. In *Proceedings of the 6th International Congress on Education in Botanic Gardens at Oxford University.* London: Botanic Gardens Conservation International.

White, J. (2019). Arboreal beings: Reading to redress plant blindness. *Australian Humanities Review*, 65.

Wolfe, C. (2010). *What is posthumanism?* Minneapolis, MA: University of Minnesota Press.

13 The Vegetal Modality of Resistance in Children's Books by/for Indigenous Peoples in the Philippines

Jose Monfred C. Sy

What primarily interests this chapter are the newest additions to the corpus of books about the Lumad. While the exonym *lumad* (lit. "native" in Cebuano) generally refers to 18 non-Muslim IP groups of the southern Philippines (Paredes, 2016, p. 332), the Lumad political identity emerged to protect their lifeways, their immediate natural environment, and their *yutang kabilin* ("ancestral land") from violations authorized by late dictator Ferdinand Marcos in the 1980s. Decades after, the Lumad movement continues to garner currency and support across the country due to their persistent and creative capacity to deal with destructive forces (Canuday, 2009, p. 1). With the help of the advocates and allied institutions, several Lumad communities established alternative schools and clinics as part of their struggle for self-determination and absolute governance over their yutang kabilin. They also take on cultural engagements such as the publication of literary works to extend the reach of their political campaigns.

Collated and published with the help of the Educators' Development Institute and the University of the Philippines Sentro ng Wikang Filipino [Center for the Filipino Language], the children's books *Ang Bayabas sa Tagaytay at Ibang Kuwento mula sa Kabataan ng Paaralang Lumad* [*The Guavas at the Ridge and Other Stories from the Youth of Lumad Schools*] and *Pangiyak: Mga Kwento at Panawagan ng mga Bayani ng Mindanao* [*Battle Cry: Stories and Calls of the Heroes of Mindanao*] introduce the Lumad lifeworld and history to (young) readers with little to no exposure to the culture of Mindanaoan non-Muslim Indigenous groups. Lumad youth can also access *Ang Bayabas* and *Pangiyak* as documents of their direct experience.

The short stories written by Indigenous youth for *Ang Bayabas* depict the sensibilities of the Lumad as farmers, underscoring values taught in Indigenous schools such as collective living at the yutang kabilin. Here, trees, ferns, and even the forest itself occupy a social status equal to humans and other members of the ecosystem. *Pangiyak* narrativizes the history of the Lumad movement by focusing on Indigenous heroes the youth could emulate the spirit of helping and service to their communities through the continued defence of local environments.

DOI: 10.4324/9781032066356-18

While there is rich scholarship on children's literature in the Philippines and Southeast Asia (Evasco, 2012; Lee, 2016; Pukan & Setiajid, 2016; Tolentino, 2008), the ecocritical potential of the genre has only recently garnered attention. Dewi (2016) and Sachmadi (2016) each survey fiction for children from across Southeast Asia as examples of "environmental literature" that promote environmental conservation and ecological literacy. Contributing to the literature on ecocritical approaches, Li & Ryan (2017) read the textual and graphic vocabularies of representing botanical life in Indonesian children's literature, foregrounding the ties between sustainability and aesthetic appeal (p. 230). In the Philippines, Noda (2018) and Lee (2019) each analyse a tropic configuration of the natural environment—wilderness/wildness and the numinous, respectively—in children's books as vectors for ecopedagogy. While these studies make sense of how Southeast Asian children's literature precipitates sustainable environmental values and practical outcomes for the betterment of humans and non-humans (Li & Ryan, 2017, p. 243), there is much room for ecocritical emphasis on children's literature written by and/or about Indigenous peoples. de Vera (2016) and Rodriguez-Tatel (2018) have already initiated studies on the representation of Indigenous children and communities in a similar set of picture books in the Philippines. This chapter makes a case for the ecological dimensions of children's books by and about the Lumad. What values and ideological operations underpin the narratives in these books? How do they reify the highly political and deeply ecological linkage between the Indigenous child and the land? In this regard, how do the books figure in the Lumad movement for self-determination?

To engage questions that traverse political, cultural, and ecological dimensions, I turn to plants. Building on budding scholarship, I contend that reflecting on, with and through plants could enrich our understanding not only of Indigenous communities, whose lifeways are ingrained to their immediate ecosystems but also of the inseparability of social justice and ecological survival. Echoing ideas on dialogue and multiplicity from Bakhtin (1984) and Deleuze and Guattari (1987), this chapter volunteers the potential of the literary text to convey that contingency to children. I argue that, narrativizing experiences of the natural environment through a relational ontology, the selected children's books about the Lumad promote a *vegetal* modality of living and resistance, one that is rooted in transcendent becoming, or coming-of-age as figured in the texts. These stories frame ethical ways of cultivating and living with plant life as crucial aspects of the Lumad movement for self-determination, whose politics is articulated as plant-like itself: rooted in land, wanting to stay in place.

The Lumad Struggle for Life and Liberation

After decades of political activism and small victories in legislation, Indigenous peoples continue to face challenges such as land grabbing, resource dispossession, displacement or forced relocation, and socio-cultural denial—forms of violence that generate profit from *both* environmental and ethnic destruction (Belisario, 2019, p. 77; Hierro & Surrallés, 2005, p. 9; IBON, 2018, p. 13). In the Philippines, the dearth of state support for basic services such as health and education exacerbates threats against the material and cultural existence of Indigenous groups (Belisario, 2019, p. 72; Lahoy & Roc, 2019, p. 12). Nevertheless, politically motivated movements, such as that of the Lumad, coalesced from terrorized Indigenous communities to commit to alternative strategies that stake their claim over their ancestral domain.

Non-Muslim ethnic groups had already settled in Mindanao even before Spain reached Philippine shores in the 1500s. Since the 16th century, colonial powers have attempted to subjugate these groups in the name of socio-political and economic expansion (Tan, 2003, pp. 25–26; Zapico et al., 2015, p. 138). The Lumad comprises the Ata, Bagobo/Ubo, Banwaon, B'laan, Kalagan, Kaulo, Dibabawon, Higaonon, Mamanwa, Mandaya, Mangguwangan, Manobo, Mansaka, Matigsalog, Subanen, Talaandig, Tiboli, and Tiruray groups. The militant Indigenous movement known today coalesced centuries after. It was on 26 June 1986 during the Lumad Mindanaw People's Federation (LMPF) founding assembly when the identity was affirmed by 18 tribal groups to lobby for self-determination, which until now are targets of the government's mineral regime (Camba, 2015, p. 8).

Some of the largest mining companies in the world operate in Mindanao, including BHP Billiton, Anglo American Corp., and Toronto Ventures Inc. These heavily extractive industries have grabbed almost half—around 700,000 hectares—of the Lumad's ancestral domain, destroying both communities and local ecologies (Canuday, 2009; SOS, 2019, p. 3). The denial of access to basic social services rendered these Indigenous communities more vulnerable to extractive and extortive enterprises. About nine out of ten Lumad did not have a proper education, and most, if not all, have no access to adequate health services (Belisario, 2019, p. 72). These challenges have prompted several communities among the Lumad to cultivate sites for educational and health services. Besides these alternative strategies, the Lumad people engage external institutions and advocates in launching public demonstrations and producing artistic and literary works (SOS, 2019).

A Room for "Radical" in Philippine Children's Literature

As vessels of protest, *Ang Bayabas* and *Pangiyak* go against the tide of commercial publication of children's literature in the Philippines. Paterno (1994) and Tolentino (2008) critique how local contemporary children's books conform to both predominant cultural expectations of childhood ("too wholesome") and commercially popular formats ("too unoriginal"). This trend is no different from the forebears of reading materials for Filipino children. During the Spanish colonization, these include prayer books and syllabaries as instructional aid in parochial schools and works we now consider "classics" imported from Europe, which are only accessible to those in higher socio-economic classes (Lee, 2016. pp. 81–82; Paterno, 1994, p. 10). The American colonial period saw an earnest growth of publication for juvenile readers. This includes Severino Reyes' *Mga Kuwento ni Lola Basyang* [*"Stories of Grandmother Basyang"*], released in instalments through *Liwayway* magazine beginning in 1925. Although hampered by the World Wars, Philippine children's literature was resuscitated by writers, artists and educators throughout the 20th century. These include figures such as Ceres O. Alabado, Amelia Lapeña-Bonifacio and Virgilio S. Almario, among others, who contributed to either production, publication, or distribution of children's literature (Lee, 2016, p. 82).

The late 1990s saw the institutional legitimization of children's literature as a literary practice in the Philippines, as evidenced by its inclusion in literary contests and anthologies (Evasco, 2012, p. 129). Several organizations such as Ilustrador ng Kabataan ("Illustrators for the Youth" [INK]) and Kuwentista ng mga Tsikiting ("Storytellers for Kids" [KUTING]) were established for the genre as many writers began taking interest in the craft. In the following years, legitimization seemed to have led to standardization. As Tolentino (2008) observes, the production of "wholesome" children's literature in quality and form—commercial, creatively illustrated, translated into English or Filipino, and containing a lesson to learn or a 'how to teach' segment at the end—is symptomatic of a consumerist movement inclined to middle class sensibilities and values (p. 1).

The publication of *Ang Bayabas* and *Pangiyak* follows the proliferation of "radical" juvenile literature and regional writing for children that began early in the 21st century (Evasco, 2012, pp. 133–134). Children's books that unsettle dominant knowledges, worldviews, and lifeways like the two challenge the "wholesome" and even conservative character of mainstream works distributed by commercial presses (p. 133). Disrupting the saturation of the ideological state apparatus, the subject matter of these books includes human rights violations marginalized in corporate media and Indigenous historical figures minoritized in traditional history textbooks. *Ang Bayabas* and *Pangiyak* contribute to the efforts of the Lumad movement for self-determination by dramatizing the communities' unshakeable conviction to defend their ancestral land and to assert their democratic rights (Pangilinan & EDI, 2019b, p. iii). As supplement to classes in Lumad schools, they serve to correct misrepresentations of Indigenous peoples in mainstream education, language, and culture (Pangilinan & EDI, 2019a, p. vi).

Thinking about, with, and through Plants

Over the past 25 years, ecocriticism has evolved in scope to encompass postcolonialism, posthumanism, animal studies, and, most recently, plants in what has been called "phytocriticism". This approach "aspires to liberate plants from the background of cultural discourse by emphasizing the vegetal dimensions of narratives and artefacts" (Li & Ryan, 2017, p. 231). Among others in the environmental humanities, Miller (2002), Nealon (2016), and Hall (2011) forage for conceptions of the vegetal across a corpus ranging from Plato to Bashō and beyond. Indigenous knowledge on (political) plant ecologies appear more commonly in anthropological studies such as those of Rose (1996; 1999), Detwiler (1992), and Grim (2001) which develop ethical models of natural stewardship disrupted by the worldwide colonial enterprise. Examples of Indigenous values confirm, inflect, and enrich proposals made in ecophilosophy about the inextricability of life and biota (its "forms").

Ecocriticism breaches the human-nature/plant dualism prevalent in "rational" Western thought, which renders the environment as a homogenized and voiceless other, ripe for exploitation (Plumwood, 1993, p. 9; Warren, 2000, pp. 46–47). Historically, this anthropocentric view embedded in cultural norms, social relationships, and value-laden judgements justified extractive regimes that contributed to climate change and losses in biodiversity (Robbins, 2012, 134). In this fictive hierarchy, plants and animals are relegated to an inferior position compared to humans due to their lack of human-like attributes such as mentality, agency, and volition (Hall, 2011, p. 1). Under the Anthropocene where human effects on the environment are more palpable than ever, "world" must be understood not as a backdrop to billion individual lives but as a territory that covers all strata of life (Nealon, 2016, p. 107). The plant, in this light, "partakes of life" (p. 30).

Exhibiting this embeddedness, Indigenous worldviews, which have suffered epistemological denial following the material violence against Indigenous communities, have been put forward as counterexamples to the Western "hyperseparation" of life (Callicott, 1997; Grim, 2001). Grim proposes that Indigenous lifeways "offer insights that may help dominant societies unlearn some things and become open to other ways of knowing the world" (p. 51), which this chapter contemplates through the figure of the plant. Local human and non-human kinship relationships demonstrated by several Indigenous groups around the world demonstrate obligations of responsibility, solidarity, and care (Hall, 2011, p. 100).

Bakhtin (1984) cultivates a model of *dialogue* that describes this kinship-based ecology. In such interactions, the "voice" or standpoint of the othered species must be heard and recognized (Bakhtin, 1984, 293). Hence, human/culture-plant/nature dialogue is only possible where there is *multiplicity*, which Deleuze and Félix Guattari explore in *A Thousand Plateaus* (1987), among other works. The vegetal figure of the *rhizome* encapsulates their theory of multiplicity:

> The multiple must be made, not always by adding a higher dimension, but rather in the simplest of ways, by dint of sobriety, with the number of dimensions that one already has available—always n−1 (the only way the one belongs to the multiple, always subtracted). Subtract the

> unique from the multiplicity to be constituted: write at n−1 dimensions. A system of this kind could be called a rhizome.
>
> (Deleuze & Félix Guattari, 1987, p. 6)

A dialogical modality of being with nature is thus moored on constant diminution—a processual *becoming*—instead of individuation, on "[growing] horizontally rather than vertically" like many vegetable species, as Miller reiterates (2002, p. 185). Nealon (2016) also notes how Deleuzoguattarean thought engages with the *vegetable soul*, "which connects everything to everything else, an ecological mesh that forms an intense territory for living" (p. 106). Multiplicity relies on the deterritorialization of subjects into a networked existence with (the) other(ed). Therefore, to foster a modality of conservation, the collective must precede the ego and operate as vitality does in the ecosystem: fluid, generous, *everywhere*.

A diversity of living Indigenous traditions demonstrate the above plant-likeness of life. In some Indigenous worldviews, plants even belong within the realm of human moral considerability (Hall, 2011, 99). The Indigenous North American Oglala, for instance, speak of the "Scared Hoop" of the Cosmos in their creation stories, where everything is bound by "wakan" (sacred power) that reflects ontological oneness (Detwiler, 1992, p. 238). Animistic societies such as the Māori are acutely aware of violating the integrity of other "persons", which include plants and animals. The trees they cut for agricultural purposes are treated as kin, and so are the tubers they eat (Hand, 1989, pp. 47–50). Here, (sustainable) consumption is permissible as long as energy flows in their kinship-based ecosystem. In Australia, Aboriginal people perceive the land or "country" as having "a will, a life force of its own" that "gives and receives life". Rose (1996) describes the Aboriginal country as a "nourishing terrain" that, as a subject on its own, knows and takes care of its people (p. 7). In these practices, life operates in a *relational space* where emergence and growth are sovereign forces (Nealon, 2016, p. 74). Thinking through plants enable an understanding of the parallel dialectical development of unity and collectivity—embraced in a process of becoming each other—emerging from Indigenous traditions. Fruitful plant-thinking informs the reading of the children's books at hand, where the recognition of relationality is followed by resistance to lifeless capitalism.

The Relational Ecology of the Vutang Kabilin

Set in relational spaces, selected stories in *Ang Bayabas* and *Pangiyak* promote alternatives to the logic of domination underlying the biospheric crisis, making them effective ecopedagogical tools that could impart the need for plant ethics (Gaard, 2009, p. 327; Li & Ryan, 2017, pp. 229–230). Any discussion on Indigenous cultures involves conflicts on land. Historically, activism has won far more legal concessions than a government's regulatory apparatuses. Conversely, Indigenous groups had to gradually integrate and adapt their institutions to the dominant legal system (Hierro & Surrallés, 2005, 10), which may exclude their punto-de-vista. By dramatizing dialogical engagements among Indigenous peoples, biota and the land, the books unsettle the state's stratification of ancestral lands through legal means of measurement and control.

In *Ang Bayabas*, the eponymous short story "Ang Bayabas sa Tagaytay" describes how the Lumad kid Paglaum guides Mona, a teacher, across the hills in a fact-finding mission against a dam construction project that will destroy parts of the forest (Ocon, 2019, p. 2). In this inaugural story and in those that followed, the organic arrangement of nature map Lumad communities: Down those hills are a group of houses. Our [settlement] ends there, and that [below the hills] is the next one (p. 3). Communities, as socio-political entities, seem to sprout from terrestrial formations where vegetation delineates borders. The story's conclusion reads:

> When [Mam Mona] looked up, she realized that they rested under a guava tree. It's full of fruit. Many of them are ripe. Mam Mona quickly picked one to quench her thirst.
>
> "Fill your pockets with guava, Paglaum!" while picking fruits.
>
> "What I ate is enough, Mam Mona. How about those who will pass by here? There won't be anything left for them."
>
> (Ocon, 2019, p. 3)

The narrative ends abruptly in the middle of their journey to prod readers to reflect on the use of plants (Hall, 2011, p. 163): Preserve (non-human) life not for yourself, but for other forms of life. This plant ethics based on care suggests a mode of environmental conservation hinged on seeking permission and offering placation (Tawhai, 2002, p. 244).

A similar care-based ethics is practised in "Bahay-Bahayan" ("Makeshift House"), which narrates how Lumad children are aerially bombed by military elements while they build a makeshift house out of vegetal excesses. Spurred by previous attacks on their homes, Doydoy and his friends pick up small branches for posts and columns for their temporary home. Someone picked up fern leaves for the roof and walls and someone else gathered *palwa ng anotong* ("dried fern stems") for the floor (Ruiz, 2019, p. 4). The children's desire for a safe refuge, imagined through and using native plants, anchors on the yutang kabilin's ecosystem. This offers a Lumad notion of territory conceived around a relational fabric that figures land as a living body that can nourish and reproduce relationships among other bodies, such as plants and children. The desire to belong to this fabric motivates characters, as in "Ang Bayabas sa Tagaytay". Other stories in *Ang Bayabas* demonstrate an analogous relational ethics practised towards animals and people. With the prominence of vegetation in the two texts, the book weaves a relational fabric where plants occupy a status equal to animals and people.

Pangiyak retells the militant activism of Lumad leaders and elders against capitalistic encroachment over their ancestral lands and immediate ecosystems. Two stories explicitly convey ethical consideration of plants Liway, the juvenile narrator of "Kalahati ng Kalangitan" ("Half of the Sky") describes the Pantaron Range—a biodiversity hotspot preyed upon by corporate entities for its mineral reserve—not only as ancestral home but as world: "For the longest time, Pantaron has been the marketplace, pharmacy, and school for the ... Manobo" (EDI, 2019b, p. 37). This narrative magnifies a scene from the life of Ba'e Bibyaon Lingkayan Bigkay, who helped form the alliance Salugpungan Ta'tanu Igkanugon ("Unity of People to Defend the Ancestral Land" in Manobo) in 1994 to unite 83 Manobo groups against the encroachment of the logging company Alcantara and Sons (AlSons) (Umil, 2018). Here and in "Ang Datu ng Talaingod" ("The Datu of Talaingod"), readers get a glimpse of the onset of the *pangayaw*, a war against AlSons and their excessive logging activities in the Lumad lifeworld.

"Ang Datu" concentrates on Datu Guibang Apoga, the Manobo leader who spearheaded the *pangayaw* with Ba'e Bibyaon ("Datu" and "Ba'e" referring to male and female Lumad leaders). Datu explains to his grandchild Manggub how the relational fabric of nature operates: "Witness the eroded soil. This was caused not by the rain but by the denudation of our forests" (EDI, 2019a, p. 66). The Datu recognizes the intentionality of trees that keep the soil intact against rain in the yutang kabilin, which then nourishes trees and prepares the Lumad's farmland. By amplifying a historical milestone in Indigenous peoples' land conflicts, the story narrativizes this "recognition of shared volition and intentionality in natural beings" (Hall, 2011, p. 106).

In these stories, Lumad characters individuate towards territory. The metaphor of the plant growth best characterizes this terrain: "Composite flowers and fruits, multiple blossoms or fruits [gather] around a single calyx" (Miller, 2002, p. 70). Narratively, they are figured as human *becomings* as they embrace a process of bringing forth an environment coded against the oppositional structure of capital and corporate extraction (Deleuze & Guattari, 1987; Nealon, 2016, p. 86). Efflorescent operations against the principle of capital vary in these narratives, ranging from the denial of greedy accumulation in "Ang Bayabas sa Tagaytay" to the recycling of wasted matter in "Bahay-Bahayan". Vitality, flowing into this rhizomatic space, "passes between organisms" (Deleuze & Guattari, 1987, p. 499).

Becoming-Trees, -Flowers and -Plants as Strategy for Resistance

Reflecting on these children's stories through the metaphor of plants further illuminates the Indigenous value of collectivity woven into textual relationalities. As ecopedagogical tools, the books position plant ethics as part of plant-like political mobilization. Marder (2013) suggests that the plant is a collective being: "its body is a non-totalizing assemblage of multiplicities, an inherently political space of conviviality" (p. 85). Ba'e Bibyaon explains to Liway in "Kalahati ng Kalangitan" a similar principle: "The whole tribe is my family ... If I married, how can I serve the tribe? My time would be occupied" (EDI, 2019b, p. 37). Collective being assumes the neuter "soul" of the plant; reproduction instantiates the multiplication of subjectivity. Like flowers in loam, such profusion is only possible in territories that act as a "nurturing receptacle", (Irigaray & Marder, 2016, p. 57) such as the relational yutang kabilin.

The two children's books demonstrate to readers how the recognition of ecological interconnection necessitates the contingencies of political-ecological survival, which many young Lumad organically learn from immediate experiences of surviving harassment and displacement from their lands. "Ang Bayabas sa Tagaytay" draws a line between the driving principles of relational life and capital accumulation (Ocon, 2019, p. 2). Against the backdrop of destructive dam construction, Paglaum privileges the generosity of conservation in consuming only what he needs. In "Bahay-Bahayan", Doydoy and his friends underscore the recreative and recuperative capacity of living beings to frustrate attempts at destruction. Narratively, this is imagined in the promise to rebuild the makeshift house destroyed by another military attack (Ruiz, 2019, p. 7). These are principles of tied community and ecology conservation.

With nourishing relationships at the core of the natural landscape, the subjectification of "setting" itself suits the anthology's logic. "Nagkakaisang Pintig" ("United Front") unfurls this potential as a site for political allegory in *Ang Bayabas*. The triumphant story describes the entry of a mining company in ancestral domain. Plants, animals, and other biotic elements comprise a political entity that unites with Lumad communities in driving the agents of capital away. "Humanity and the environment are a united front", imagines the story, "but soon enough, with the growth of cities, humans realized that generosity has an opposite—greed". Two arguments are postulated here: Firstly, the natural environment, which includes all forms of life, immanently regulate the flow of life to sustain all species; and secondly, it is human greed, which is *unnatural*, that split humanity from naturality, a nod to the colonial enterprise that first disturbed the Lumad lifeworld centuries ago. This story restitutes to the vegetal members of the forest, as much as to humans, their "right to exist, to grow, and to flourish" (Marder, 2012, p. 31).

To dramatize how the Lumad view their resistance, the narrative fictionalizes the Konseho ng Kalikasan ("Council of the Environment"), led by the forest with the assistance of Pag-asa ("hope") the eagle, which organizes other natural beings into mobilizing with the Lumad (p. 31). Subjectivity, in the form of agency, is part of all forms and sites of life (Rose, 1999, p. 178), as this story punctuates. Three days later, the Council joined the Indigenous community in their protest and prompted a landslide over the mining company. "Defend the land", they collectively chant (p. 32). A moral consideration promoting conservation blooms from the multiplicity of actors in this dialogical relationship (Bakhtin, 1984, p. 293). While the landslide could be read as a consequence of deforestation, the story codes interspecies collectivity by endowing the forest and other biota with intentionality.

Through inter-generational narratives, *Pangiyak* reminds us that this vegetal modality of resistance has existed as real-life praxis. In the parallel historical accounts of Ba'e Bibyaon and Datu Guibang, the united Manobo groups staged a war against AlSons as an attempt to preserve the natural environment and the Indigenous communities (EDI, 2019, p. 40). Ba'e Bibyaon's narrative ends with a ritual dance that inaugurates this battle (p. 42). Ritual reenactment deepens the kinship ecology in the yutang kabilin by brushing off humans' "false sense of themselves as superior" (Hall, 2011, p. 162). The term "vegetal" best describes this radical activity that hinges on the desire for continued growth. After all, any progressive dialectic "must consistently produce ever more new, indifferent, asignifying remainders", (Nealon, 2016, p. 69) or the Lumad themselves who exceed the state's biopolitics. All forms of nature play a part in the relational fabric. As Datu Guibang intimates to his grandson, the absence of trees warrants the erosion of soil; parallelly, the displacement of Lumad communities promise the dissolution of the yutang kabilin. Imparting plant ethics to the young, the older generation frame the environment as not at all other to the Lumad. Staking the Lumad's claim on their land constitutes a visible extension—a rootedness—without interiority (Hegel, 2004, p. 308). Signified by rhizomatic growth, the individual becomes territory, in the sense of "nourishing terrain" (Rose, 1996) and vice versa.

These stories from the two children's books suggest that, for Indigenous peoples, plants do not only figure as tropes of political activism; they also resist capitalistic encroachment in their very material presence. While *purely* vegetal beings do not protest, humans could ground resistance in a praxial fiction of vegetality to cultivate a non-appropriative way of resonating with and protecting the natural environment (Marder, 2012, p. 26). After all, a "movement" does not only operate on mobility but also on growth (Aristotle in Marder, 2012, p. 28) as shown by the importance of local ecologies for the Lumad people.

Conclusion: Conservation as Protest

By featuring how Lumad youth think and live on land, readers are invited to an ancestral domain constructed out "of a very specific and singular fabric of social ties between the different beings that make up that environment" (Hierro & Surrallés, 2005, 11). The stories in *Ang Bayabas sa Tagaytay* and *Pangiyak* champion the relational fabric that allows for multivocal dialogue, a site for cultivating ethical and moral consideration (Baxter & Montgomery, 1996, p. 24). Life according to these proposals is cast as vegetal in its collectivity—disseminating and productive and not hierarchical and reductive as cemented by a Western, rationalist dualism (Miller, 2002, p. 183). Indeed, "intergenerational, inter-cultural, and inter-species communities" could generate narratives of social and environmental justice (Gaard, 2009, p. 334). Whatever grows from this land carries the seeds of political potential.

That the two *children*'s books are auspicious for a plant-thinking project is no accident. Published for an imagined child, the genre is a cultural sign of "adult political standards and socialization practices" (May, 1997, p. 82). Against a tide of "wholesome" and "unoriginal" production, *Ang Bayabas* and *Pangiyak* impart radical ethics and politics—practised on ground by the Lumad movement—to children. Engaged with a political campaign, the books hope for continuity, embodied as both children and plants. Juvenile readers, in a processual coming-of-age themselves, are thus involved in the stories' inscription of conservation into a modality of living and resistance—non-totalizing, collective, and always in a state of becoming, as children growing up are.

This "militant spirit and care for humans and nature", Datu Guibang explains, must be preserved for the next generations. Childhood is a time ripe for "developing individual potential suited to a future in which societies could be different in some significant ways" (Reynolds, 2007, p. 2). Children, after all, are sites of much ideological activity. Embracing the politics of mutual responsibility, the Indigenous knowledge narrativized in these children's books by and for the Lumad decidedly invites readers young and old to return to nature by conserving it.

References

Bakhtin, Mikhail. (1984). *Problems of Dostoevsky's poetics*. Minneapolis, MN: University of Minnesota Press.

Baxter, L. A., & Montgomery, B. M. (1996). *Relating: Dialogues and dialectics*. New York: Guilford Press.

Belisario, Paul. (2019). Lumad schools: A struggle for land and learning. In J. del Rosario-Malonzo (Ed.), *Repression and impunity: Continuing attacks on people's rights in the Philippines* (pp. 71–80). Quezon City: IBON International.

Callicott, J. B. (1997). *Earth's insights: A multicultural survey of ecological ethics from the Mediterranean basin to the Australian outback*. Berkeley, CA: University of California Press.

Camba, A. A. (2015). From colonialism to neoliberalism: Critical reflections on Philippine mining in the "long twentieth century". *The Extractive Industries and Society*, 2(2), 287–301.

Canuday, J. J. (2009). *Bakwit: The power of the displaced*. Quezon City: Ateneo de Manila University Press.

de Vera, C. P. (2016). Close reading of selected titles about Indigenous children. In Elisabeth Oseanita Pukan & Haris HermansyahSetiajid (Eds.), *Proceedings of the 4th literary studies conference: Children's literature in Southeast Asia* (pp. 59–65). Yogyakarta, Indonesia: Universitas Sanata Dharma.

Deleuze, G. & Guattari, F. (1987). *A thousand plateaus: Capitalism and schizophrenia* (Brian Massumi, Trans.). Minneapolis: University of Minnesota Press.

Detwiler, F. (1992). All my relatives: Persons in Oglala religion. *Religion*, 22, 235–246.

Dewi, N. (2016). Of people and nature: Eco education in young adult literature from Asia. In Elisabeth Oseanita Pukan & Haris HermansyahSetiajid (Eds.), *Proceedings of the 4th literary studies conference: Children's literature in Southeast Asia* (pp. 254–259). Yogyakarta, Indonesia: Universitas Sanata Dharma.

Echeverri, J. Á. (2005). Territory as body and territory as nature: Intercultural dialogue? In P. Hierro & A. Surrallés (Eds.), *The land within: Indigenous territory and perception of environment* (pp. 230–247). Copenhagen: International Work Group for Indigenous Affairs.

Educators' Development Institute (EDI). (2019a). Ang Datu ng Talaingod. In S. Pangilinan & Educators' Development Institute (Eds.), *Pangiyak: Mga Kuwento at Panawagan ng mga Bayani ng Mindanao* (pp. 61–70). Quezon City: Sentro ng Wikang Filipino.

Educators' Development Institute (EDI). (2019b). Kalahati ng Kalangitan. In S. Pangilinan & Educators' Development Institute (Eds.), *Pangiyak: Mga Kuwento at Panawagan ng mga Bayani ng Mindanao* (pp. 33–42). Quezon City: Sentro ng Wikang Filipino.

Evasco, E. (2012). Ang Ikatlong Dekada: Ang Sitwasyon at Mungkahing Pagpapaunlad ng Panitikang Pambata sa Pilipinas (2000–2009). *Daluyan: Journal ng Wikang Filipino*, 18(1–2),128–162.

Gaard, G. (2009). Children's environmental literature: From ecocriticism to ecopedagogy. *Neohelicon* 36(2), 321–334.

Grim, J. (Ed.). (2001). *Indigenous traditions and ecology*. Cambridge, MA: Harvard University Press.

Hall, M. (2011). *Plants as persons: A philosophical botany*. New York: State University of New York Press.

Hand, S. (Ed.). (1989). *The Levinas reader*. Oxford: Blackwell.

Hegel, G. W. F. (2004). *Philosophy of nature: Encyclopedia of the philosophical sciences, part II* (A. V. Miller, Trans.). Oxford: Oxford University Press.

Hierro, P. G, & Surrallés, A. (2005). Introduction. In P. Hierro & A. Surrallés (Eds.), *The land within: Indigenous territory and perception of environment* (pp. 8–10). Copenhagen: International Work Group for Indigenous Affairs.

IBON Institute for International Development (IBON). (2018). *Simmering crisis, rising inequalities amid corporate concentration: International situation*. Quezon City: IBON International.

Irigaray, L., & Marder, M. (2016). *Through vegetal being: Two philosophical perspectives*. New York: Columbia University Press.

Lahoy, R. Jr., & Roc, R. (2019). A creeping reign of repression, impunity in the Philippines. In J. del Rosario-Malonzo (Ed.), *Repression and impunity: Continuing attacks on people's rights in the Philippines* (pp. 1–16). Quezon City: IBON International.

Lee, G. (2016). Mix and match: A thematic analysis of the depictions of alternative family arrangements in Philippine children's picturebooks. *Journal of English Studies and Comparative Literature*, 15(1), 77–112.

Lee, G. (2019). Return to innocence: Encountering the numinous in children's fantasy fiction by Cyan Abad-Jugo. In Chi P.Pham, ChitraSankaran, & Gurpreet Kaur (Eds.), *Ecologies in Southeast Asian literatures: Histories, myths and societies* (pp. 49–57). Malaga: Vernon Press.

Li, R. & Ryan, J. C. (2017). Sowing seeds: Phytocriticism and the botanical dimensions of Indonesian literature for children and young adults. In John Charles Ryan (Ed.), *Southeast Asian ecocriticism: Theories, practices, prospects* (pp. 229–251). Lanham, MD: Lexington Books.

Marder, M. (2012). Resist like a plant! On the vegetal life of political movements. *Peace Studies Journal*, *5*(1), 24–32.

Marder, M. (2013). *Plant-thinking: A philosophy of vegetal life*. New York: Columbia University Press.

May, J. (1997). Theory and textual interpretation: Children's literature and literary criticism. *The Journal of the Midwest Modern Language Association*, 30(1/2), 81–96.

Miller, E. (2002). *The vegetative soul: From philosophy of nature to subjectivity in the feminine*. New York: State University of New York Press.

Nealon, J. (2016). *Plant theory: Biopower and vegetable life*. Stanford: Stanford University Press.

Noda, A. A. (2018). Reconstructing the wilderness: Finding identity, culture, and values in Filipino children's literature. *The Journal of Ecocriticism*, 8(1), 1–8.

Ocon, H. (2019). Ang Bayabas sa Tagaytay. In S. Pangilinan & Educators' Development Institute (Eds.), *Ang Bayabas sa Tagaytay at Iba pang Kuwento mula sa Kabataan ng Paaralang Lumad* (pp. 2–3). Quezon City: Sentro ng Wikang Filipino.

Pangilinan, S. & Educators' Development Institute. (2019a). *Ang Bayabas sa Tagaytay at Iba pang Kuwento mula sa Kabataan ng Paaralang Lumad*. Quezon City: Sentro ng Wikang Filipino.

Pangilinan, S. & Educators' Development Institute. (2019b). *Pangiyak: Mga Kuwento at Panawagan ng mga Bayani ng Mindanao*. Quezon City: Sentro ng Wikang Filipino.

Paredes, O. (2016). Rivers of memory and oceans of difference in the Lumad world of Mindanao. *TRaNS: Trans-Regional and National Studies of Southeast Asia*, 4(2), 329–349.

Paterno, M. E. (1994). A history of children's literature in the Philippines. In Virgilio S.Almario, MariaElena Paterno, Ramon C.Sunico, & Rene O.Villanueva (Eds.), *Bumasa at Lumaya: A sourcebook on children's literature in the Philippines* (pp. 9–23). Pasig City: Anvil Publishing.

Plumwood, V. (1993). *Feminism and the mastery of nature*. London: Routledge.

Pukan, Elisabeth Oseanita & Haris Hermansyah Setiajid (Eds.). (2016). *Proceedings of the 4h literary studies conference: Children's literature in Southeast Asia*. Yogyakarta, Indonesia: Universitas Sanata Dharma.

Reynolds, Kimberley. (2007). *Radical children's literature future visions and aesthetic transformations in juvenile fiction*. New York: Palgrave Macmillan.

Robbins, Paul. (2012). *Political ecology* (2nd ed.). Chichester: Wiley-Blackwell.

Rodriguez-Tatel, Mary Jane. (2018). Imaging the "indigenous"child in selected stories for children: Revisiting and revisioning "tribe" as a colonial discourse. *Humanities Diliman*, 15(1), 57–84.

Rose, Deborah. (1996). *Nourishing terrains: Australian Aboriginal views of landscape and wilderness*. Canberra: Australian Heritage Commission.

Rose, Deborah. (1999). Indigenous ecologies and an ethic of connection. In N. Low (Ed.), *Global ethics and environment* (pp. 175–187). London: Routledge.

Ruiz, Elmo. (2019). Bahay-bahayan. In S. Pangilinan & Educators' Development Institute (Eds.), *Ang Bayabas sa Tagaytay at Iba pang Kuwento mula sa Kabataan ng Paaralang Lumad* (pp. 4–7). Quezon City: Sentro ng Wikang Filipino.

Sachmadi, Ida Farida. (2016). Literacy on health and environment in eight novels written by Indonesian children. In Elisabeth Oseanita Pukan & Haris HermansyahSetiajid (Eds.), *Proceedings of the 4th literary studies conference: Children's literature in Southeast Asia* (pp. 126–129). Yogyakarta, Indonesia: Universitas Sanata Dharma.

Sambil, Ma. Teresa. (2019). Nagkakaisang Pintig. In S. Pangilinan & Educators' Development Institute (Eds.), *Ang Bayabas sa Tagaytay at Iba pang Kuwento mula sa Kabataan ng Paaralang Lumad* (pp. 30–33). Quezon City: Sentro ng Wikang Filipino.

Save Our Schools Network (SOS). (2019). *Intensifying the people's struggle amidst state fascism* [Primer].

Tan, Samuel. (2003). *Internationalization of the Bangsamoro struggle*. Quezon City: UP Center for Integrative and Development Studies.

Tawhai, T. P. (2002). Maori religion. In G. Harvey (Ed.), *Readings in indigenous religions* (pp. 237–249). London: Cassell Academic.

Tolentino, Rolando. (2008). Pinag-aagawang Bata sa Panitikang Pambata: Folklore, media at Diskurso ng Bata. *Malay*, 21(1), 1–11.

Umil, Anne Marxze. (2018, Dec. 1). Context of Talaingod incident: The decades-old struggle of Lumad in Pantaron Mountain range for ancestral land, right to self-determination. Retrieved from https://www.bulatlat.com/.

Warren, Karen. (2000). *Ecofeminist philosophy*. Lanham, MD: Rowman & Littlefield.

Zapico, F. L., C. H. Aguilar, A. Abistano, J. C. Turner, & L. J. Reyes. (2015). Bio-cultural diversity of Sarangani Province, Philippines: An ethno-ecological analysis. *Rice Science*, 22(3), 138–146.

Index

Kolibri 360 GmbH
Ferdinand-Jühlke-Straße 7
99095 Erfurt, Deutschland
produktsicherheit@kolibri360.de